Synthesis, Properties, and Crystal Chemistry of Perovskite-Based Materials

Technical Resources

Journal of the American Ceramic Society

www.ceramicjournal.org

With the highest impact factor of any ceramics-specific journal, the *Journal of the American Ceramic Society* is the world's leading source of published research in ceramics and related materials sciences.

Contents include ceramic processing science; electric and dielectic properties; mechanical, thermal and chemical properties; microstructure and phase equilibria; and much more.

Journal of the American Ceramic Society is abstracted/indexed in Chemical Abstracts, Ceramic Abstracts, Cambridge Scientific, ISI's Web of Science, Science Citation Index, Chemistry Citation Index, Materials Science Citation Index, Reaction Citation Index, Current Contents/ Physical, Chemical and Earth Sciences, Current Contents/Engineering, Computing and Technology, plus more.

View abstracts of all content from 1997 through the current issue at no charge at www.ceramicjournal.org. Subscribers receive full-text access to online content.

Published monthly in print and online. Annual subscription runs from January through December. ISSN 0002-7820

International Journal of Applied Ceramic Technology

www.ceramics.org/act

Launched in January 2004, *International Journal of Applied Ceramic Technology* is a must read for engineers, scientists,and companies using or exploring the use of engineered ceramics in product and commercial applications.

Led by an editorial board of experts from industry, government and universities, *International Journal of Applied Ceramic Technology* is a peer-reviewed publication that provides the latest information on fuel cells, nanotechnology, ceramic armor, thermal and environmental barrier coatings, functional materials, ceramic matrix composites, biomaterials, and other cutting-edge topics.

Go to www.ceramics.org/act to see the current issue's table of contents listing state-of-the-art coverage of important topics by internationally recognized leaders.

Published quarterly. Annual subscription runs from January through December.
ISSN 1546-542X

American Ceramic Society Bulletin

www.ceramicbulletin.org

The *American Ceramic Society Bulletin*, is a must-read publication devoted to current and emerging developments in materials, manufacturing processes, instrumentation, equipment, and systems impacting the global ceramics and glass industries.

The *Bulletin* is written primarily for key specifiers of products and services: researchers, engineers, other technical personnel and corporate managers involved in the research, development and manufacture of ceramic and glass products. Membership in The American Ceramic Society includes a subscription to the *Bulletin*, including online access.

Published monthly in print and online, the December issue includes the annual *ceramicSOURCE* company directory and buyer's guide. ISSN 0002-7812

Ceramic Engineering and Science Proceedings (CESP)

www.ceramics.org/cesp

Practical and effective solutions for manufacturing and processing issues are offered by industry experts. CESP includes five issues per year: Glass Problems, Whitewares & Materials, Advanced Ceramics and Composites, Porcelain Enamel. Annual subscription runs from January to December.
ISSN 0196-6219

ACerS-NIST Phase Equilibria Diagrams CD-ROM Database Version 3.0

www.ceramics.org/phasecd

The ACerS-NIST Phase Equilibria Diagrams CD-ROM Database Version 3.0 contains more than 19,000 diagrams previously published in 20 phase volumes produced as part of the ACerS-NIST Phase Equilibria Diagrams Program: Volumes I through XIII; Annuals 91, 92 and 93; High Tc Superconductors I & II; Zirconium & Zirconia Systems; and Electronic Ceramics I. The CD-ROM includes full commentaries and interactive capabilities.

Synthesis, Properties, and Crystal Chemistry of Perovskite-Based Materials

Ceramic Transactions Volume 169

Proceedings of the 106th Annual Meeting of The American Ceramic Society, Indianapolis, Indiana, USA (2004)

Editors

Winnie Wong-Ng
Amit Goyal
Ruyan Guo
Amar S. Bhalla

Published by
The American Ceramic Society
PO Box 6136
Westerville, Ohio 43086-6136
www.ceramics.org

For information on ordering titles published by The American Ceramic Society, or to request a publications catalog, please call 614-794-5890, or visit our website at www.ceramics.org

ISBN 1-57498-190-0

Contents

Crystal Chemical and Physical Related Issues, Synthesis and Crystal Growth

Preface

Ceramic materials with perovskite-type structure encompass a large number of chemical compounds that are of considerable technological importance due to their rich crystal chemistry. The applications of perovskites include piezoelectric transducers, multilayer capacitors, PTC thermistors, electrooptical modulators, optical switches, dielectric resonators, thick film resistors, electronic sensors, electrorestrictive actuators, magnetic bubble memory devices, laser host materials, ferromagnetic materials, refractory electrodes, second harmonic generators, batteries, ceramic electrodes, thermoelectric devices, and high temperature superconductors. In each of these applications, both the intrinsic as well as the extrinsic issues are important in determining the properties of the device in question.

This proceedings volume contains articles that were presented at the Electronics Division Focused Session on Synthesis, Properties, and Crystal Chemistry of Perovskite-Based Materials during the 106th Annual Meeting of The American Ceramic Society (ACerS), April 18-21, 2004 in Indianapolis, Indiana. This focused session has brought together experts in different areas of perovskite research to address both the intrinsic and extrinsic issues. In the area of intrinsic issues, papers were presented in research or development of new perovskite materials for various applications, including doping of existing perovskite materials. In the area of extrinsic issues, papers concerned processing routes, which result in improved properties. Different processing routes can result in materials with different grain sizes, modified grain boundary misorientation distributions, modified crystallographic textures, different densities, etc.

The articles in this volume are divided into three parts: novel applications of perovskite materials; processing, texture and grain boundaries in perovskites; and crystal chemical and physical related issues, synthesis and crystal growth. The order in which the papers appear here and the division into which they are organized may be different from that of their presentation at the meeting. It is hoped that this volume will provide a comprehensive summary of the latest development in perovskite research and will make available source material for researchers and managers working in this field.

We acknowledge Mr. Greg Geiger and Mr. Bill Jones for their involvement in editing and producing this book. Special thanks are due to the speakers, authors, symposium session chairs, manuscript reviewers, and ACerS program coordinators and officials for their contributions. Without their efforts, this symposium and proceedings would not have been a successful.

Winnie Wong-Ng
Amit Goyal
Ruyan Guo
Amar S. Bhalla

Novel Applications of Perovskite Materials

BISMUTH-BASED PEROVSKITE STRUCTURE SOLID SOLUTIONS WITH FERROELECTRIC MORPHOTROPIC PHASE BOUNDARIES FOR PIEZOELECTRIC APPLICATIONS

L. Eric Cross and Jinrong Cheng
187 Materials Research Institute
The Pennsylvania State University
University Park, Pa 16802 USA

ABSTRACT

This article summarizes a range of studies exploring solid solutions in the $BiBO_3$:$PbTiO_3$ family, where the B site cations are chosen to enhance the ferroelectric Curie Temperature with the objective of expanding the temperature range of morphotropy and the corresponding range of high piezoelectric properties. Recent studies have focused upon $BiFeO_3$ - based compositions as the ferrite has the highest Curie Point for any known perovskite. Heretofore ceramics in this family have been plagued by low electrical resistivity, rendering them useless for piezoelectricity. This work has shown that minor additions of gallium can enhance the resistivity up to 10^{13} ohm cm, providing ceramics with excellent high field dielectric properties. Even for compositions close to the Morphotropic Phase Boundary (MPB) the high coercivity of the ferrite solid solution makes them difficult to pole and it has been necessary to sacrifice some Curie range in the ceramic by lanthanum addition to achieve polability and piezoelectric properties. Parallel work at Penn State (See the chapter by Randall et al.) has focused upon $BiScO_3$:$PbTiO_3$: and shown MPB composition ceramics closely comparable in properties to PZT. Further it has been shown that single crystals of the MPB compositions can be grown and cut to yield d_{33} ~1,250 pC/N and k_{33} ~ 90% up to 300° C. In the ferrite compositions under study, the next step will be to grow single crystals. It is important to note that even at the MPB the tetragonal compositions have c/a larger than 1.05, promising large switchable strain.

INTRODUCTION

In ferroelectric material systems which will be of importance for practical uses as strong piezoelectrics in applications such as actuation or power transduction some of the following attributes will be essential. High dielectric permittivity coupled with strong spontaneous electric polarization, implying a high ferroelectric Curie temperature together with high paraelectric prototypic symmetry. Large permittivity in a strongly polar state implies proximity to a ferroelectric: ferroelectric phase transition, and to retain such proximity over a wide temperature range mandates the engineering of a solid solution system which incorporates a morphotropic phase boundary (MPB) in a part of the phase space accessible by conventional processing either for ceramic forming or single crystal growth. The material must be of a good insulator with high dielectric breakdown strength to permit poling and actuation at high field levels.

This brief review will summarize several years of work in the Materials Research Laboratory/Institute at Penn State exploring bismuth based perovskite structure ferroelectrics as end members in solid solutions with lead titanate. The venture was initially very high risk for two reasons. Firstly, the Bi^{3+} cation is too small to comfortably occupy the A site in ABO_3 perovskite. If formed they would be expected to be rhombohedral ferroelectrics with high Curie point. Several can be stabilized in solid solution with lead titanate ($PbTiO_3$) and the risky question is starting from the $PbTiO_3$ end of the solid solution which is surely stable can one achieve the tetragonal:rhombohedral (MPB) before the loss of structural stability. A system where success was achieved is the $BiScO_3$:$PbTiO_3$ solid solution where the MPB composition can just be achieved. In $BiGaO_3$:$PbTiO_3$ in spite of intense efforts the MPB could not be realized, and it was necessary to move to $BiGa_xSc_{1-x}O_3$:$PbTiO_3$ to achieve a stable boundary.

A system where the problem does not obtrude is $BiFeO_3$:$PbTiO_3$ solid solution. Bismuth ferrate is a stable perovskite with Tc at 850° C and thus the expectation of massive Ps. However now a second problem evidences, it is that of achieving and maintaining high insulation resistance against the easy valence change Fe^{3+} to Fe^{2+}. A useful accidental consequence of the gallate program was the finding that small additions of $BiGaO_3$ had a profound effect in the raising the missing breakdown strength of $BiFeO_3$ and associated solid solutions.

EXPERIMENTAL PROCEDURE

The crystalline solid solutions explored were largely fabricated using convention mixed oxide ceramic processing. The starting material were Bi_2O_3, Ga_2O_3, Sc_2O_3, Fe_2O_3, $PbCO_3$ and TiO_2, all with purities greater than 99%. For each composition of the different solid solutions suitable fractions of the appropriate components were mixed by ball milling for 24 hours with stabilized ZrO_2 media, then calcined at 750°–850°C for 2 to 4 hours. After Calcining the powders were vibratory milled for an additional 24 hours to ensure uniformity, then uniaxially pressed into pellets with 12.7 mm diameter and 1.5 mm thickness. Final forming was in a sealed crucible under protective atmosphere in the temperature range from 950° to 1100°C for times of order 0.8 to 1.5 hours according to composition. The sintered weight loss and radial shrinkage were of order 1% and ~ 18% respectively.

Structure and phase evolution as a function of composition and temperature were characterized by X-ray diffraction (XRD) (Sintag-1). Microstructure was explored using freshly fractured surfaces of sintered pellets examined by scanning electron microscopy (SEM) (Hitachi S-3000H).

For electrical studies polished disks were annealed then electroded with a post fire silver paste (Dupont 6160) Temperature dependent weak field dielectric measurements were made using a controlled HP-4284 impedance analyzer with the sample housed in a Delta Design environmental control chamber. Electrical insulation resistance was measured using a Keithley 6517 meter. The ferroelectric hysteresis and induced elastic strain were measured simultaneously using a modified Sawyer Tower setup with strain measurement by a linear variable differential transformer. Piezoelectrics properties were characterized by Berlincourt d_{33} meter and checked by the IEEE resonance-antiresonance method with an HP-4194A impedance analyzer.

BISMUTH GALLATE:LEAD TITANTE SOLUTIONS

In spite of using every stratagem consistent with conventional ceramic processing it was not possible to produce a stable single phase perovskite structure in solid solution with more than 25% $BiGaO_3$. [(1)] At this composition the structures is still clearly tetragonal (Fig.1) with no evidence of a rhombohedral ferroelectric phase. As expected both differential thermal analysis Fig. 2a and dielectric permittivity measurements (Fig. 2b) show evidence of a ferroelectric phase change in the vicinity of 490°C confirming the high Curie temperatures in this solid solution system. It is probable that compositions up to and including the morphotropic phase boundary could be stabilized under high pressure conditions, but this was beyond the scope of the present program.

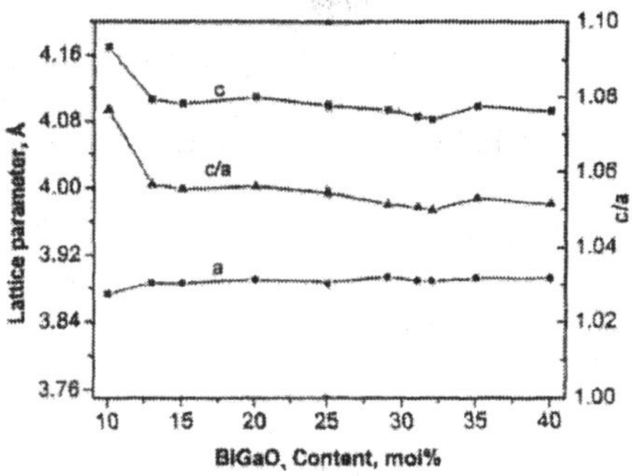

Fig. 1: Lattice constants of BG-Pt as the function of GB content

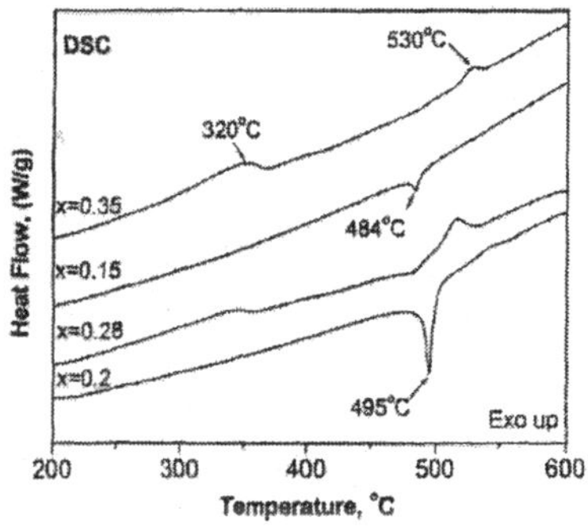

Fig 2a: DSC curves of BGPT ceramic powders with various compositions

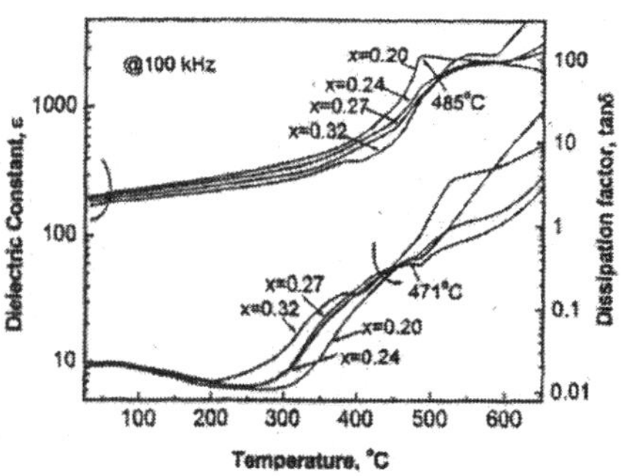

Fig. 2b: The temperature dependent dielectric spectrum taken at 100 kHz.

BISMUTH SCANDATE: LEAD TITANATE SOLID SOLUTIONS

A more extensive sequence of studies of the $BiScO_3$:$PbTiO_3$ solid solution systems has been carried forward in the MRI at Penn State, and in a joint program with Dr. T. Sekiya's group at NIRIN in Nagoya, now part of the Smart Structures Research Center at NAIST in Tsukuba. Evidence is that by careful processing in sealed atmosphere controlled crucibles.[(2)] [(3)] It was possible to produce single phase perovskite structure ceramics up to 47% $BiScO_3$ and that a tetragonal rhombohedral boundary is clearly evident at 37.5% $BiScO_3$. (Fig. 3) Unlike PZT the Curie Temperature first rises with $BiScO_3$ addition to $PbTiO_3$ and is over 400°C at the MPB composition (Fig 4.a) leading to very high values of the remanent polarization (Fig. 4b) and large values of d_{33} comparable to soft PZT. (Fig. 5)

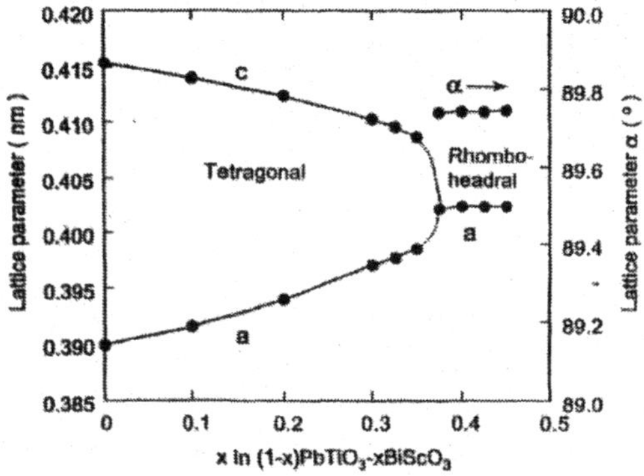

Fig. 3: Plot of lattice parameters vs. composition for $(1-x)$PT-xBS perovskites

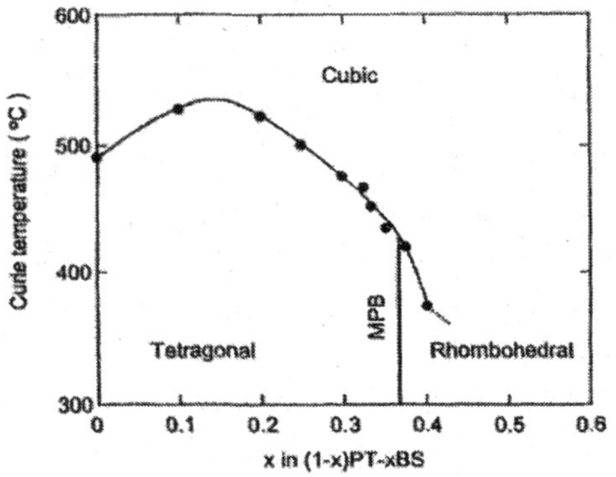

Fig. 4a: Phase diagram in the PT-BS system

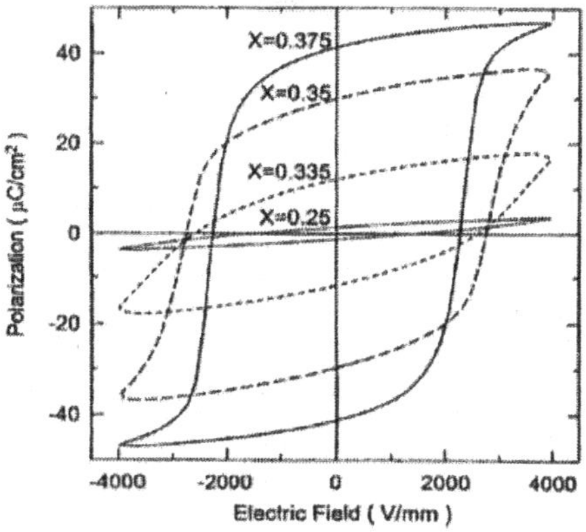

Fig. 4b: D-E hysteresis loop of $(1-x)$ PT-xBS ceramics at room temperature

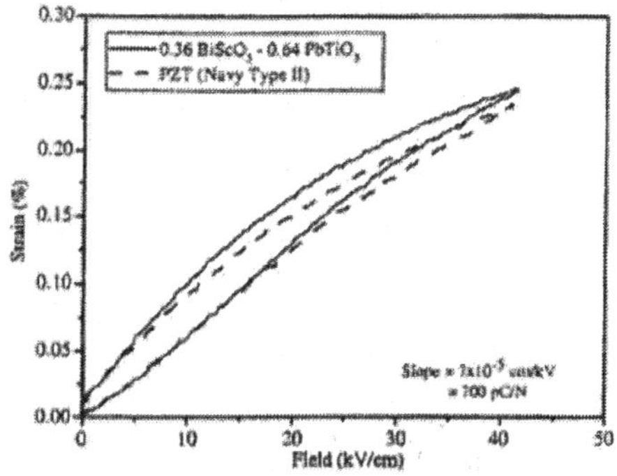

Fig. 5: Unipolar strain-electric field behavior of poled 36% $BiScO_3$ -64% $PbTiO_3$, contrasted to a Navy Type II PZT, showing a high field d_{33} of 700 pC/N

Clearly a cogent reason for exploring alternative MPB systems with high Curie Temperature and corresponding large polarization is that unlike PZT they may be tractable for single crystal growth and thus offered opportunity to explore oriented properties, realizing ultra high values of d_{33} and k_{33} as in the 001 poled PZN:PT and PMN:PT crystals. Recent studies at Penn State [(4)] have shown that indeed it is possible to grow single crystals at the 43% $BiScO_3$ composition, that 001 oriented specimens do have very high d_{33} ~ 1,200 pC/N and coupling coefficient k_{33} ~ 90% maintained to over 330° C (Fig. 6a) with strain capability over 0.12% at 10kV/cm corresponding to the weak field d_{33} ~ 1,200 pC/n (Fig. 6b). The single crystal BS:PT data is of critical importance as it does suggest that the polarization tilting mechanism which gives the ultra high response in PZN:PT is general and likely to be possible in other high Curie Temperature solid solution systems with MPBs.

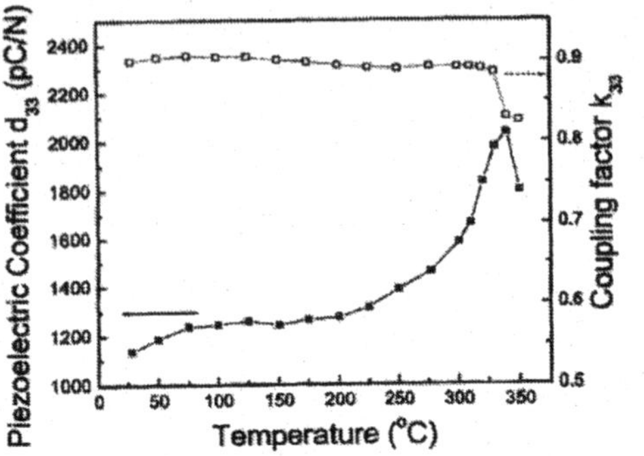

Fig. 6a: Longitudinal electromechanical coupling factors k_{33} and piezoelectric coefficients d_{33} as a function of temperature

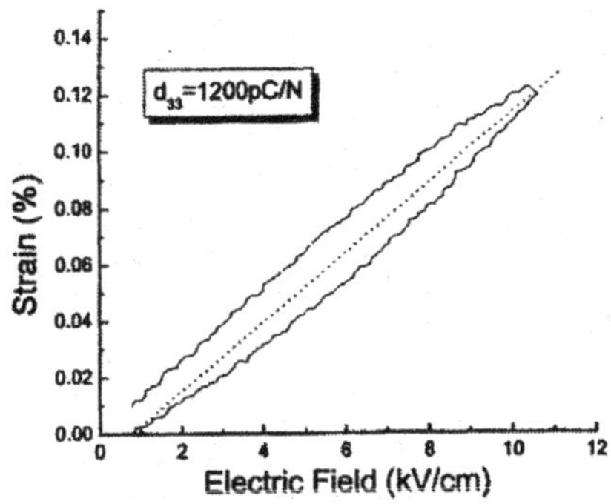

Fig. 6b: Unipolar S-E loop for <001) –oriented BSPT (43/57) single crystals

BISMUTH SCANDATE GALLATE:LEAD TITANATE SOLID SOLUTION

The success achieved with the bismuth scandate based solid solutions suggests that it should be interesting to explore gallate:scandate combination where the scandium may stabilize the solid solution with lead titanate sufficiently to achieve the MPB phase boundary compositions. Initial studies using (1-x)Bc $(Ga_{1/4}Sc_{3/4})O_3$-x$PbTiO_3$ suggests that this may be the limit for gallium substitution. The potential advantage for the mixed systems is shown in Fig. 7 where it is evident that the Curie Temperature is raised into the range 465 to 510°C however, this is at the cost of some second phase contamination in the rhombohedral phase compositions below 60% $PbTiO_3$.[5] It will be interesting to see how minor additions of $BiGeO_3$ effect crystal growth of $BiScO_3$ crystals and if the addition has the same effect on conduction and dielectric strength.

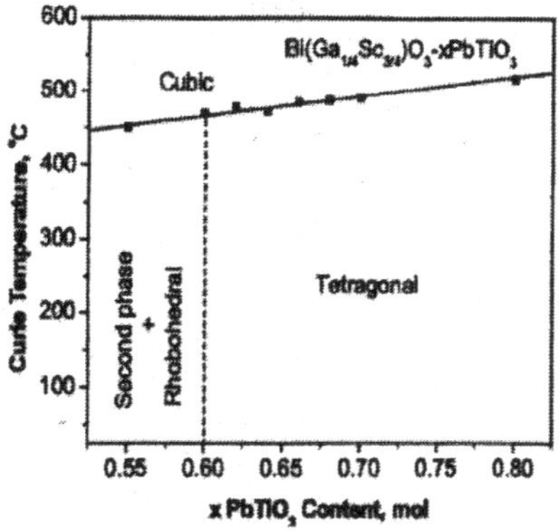

Fig. 7: Draft of a phase diagram of BGS-PT as a function of the PT content

BISMUTH FERRATE: LEAD TITANATE BASED SOLID SOLUTIONS:

Introduction

Bismuth Ferrate $BiFeO_3$ is an obvious end member candidate for solid solution with $PbTiO_3$. It is quite stable in the perovskite form with a Curie point at 850° C [(6)] and an antiferromagnetic Néel temperature at 310°C. [(7)] The first Penn State studies focused upon exploring "dopants" to control conduction. This work showed that Ga^{3+} additions could generate resistivities up to 10^{13} ohm cm in compositions of $BiFeO_3$:$PbTiO_3$ close to the MPB (near 0.3 mole $PbTiO_3$). Not only were these ceramics of very high resitivity, but they also had excellent breakdown characteristics, supporting fields up to 250 kV/cm.[(8)] Even at these very high field levels it proved difficult to pole samples of ceramic even at compositions near the MPB so a range of lanthanum modified compositions was explored, where some of the high Tc and high strain were sacrificed but ceramics poling could be achieved.

Composition chosen for initial study

In gallium modified $Bi(Ga_xFe_{1-x})O_3$:$PbTiO_3$ compositions chosen for initial study were 0.7 BGF:0.3PT for x in the range $0 \leq x \leq 0.1$ and 0.4 BGF:0.6PT for x in the range of $0 \leq x \leq 0.5$. For the lanthanum modified compositions, studies were focused upon (1-x) $(Bi_{1-y}La_{y)}$ $(Ga_{0.05}Fe_{0.95})O_3$: x $PbTiO_3$ (BLGF:PT) for x values in the range $0.2 \leq x \leq 0.65$ and for y values in the range $0.2 \leq x \leq 0.65$ and for y values in the range $0 \leq y \leq 0.20$. Earlier studies had shown that the level 0.05 Ga was adequate to maintain high insulation resistance in the interesting MPB compositions.

Characterization

The effects of Ga substitution on the phase evolution in $Bi(Ga_xFe_{1-x})O_3$:$PbTiO_3$ (BGF:PT) ceramics were identified by XRD analysis. Figures 8 (a) and (b) show the diffraction patterns for 0.7 BGF: 0.3PT with Ga concentration $x \leq 0.10$ and for 0.4 BGF:0.6 PT with x in the range $x \leq 0.5$.

Coexistence of rhombohedral and tetragonal phases was observed for 0.7 BGF:0.3PT as marked in Fig. 8(a) showing that the 0.3PT composition lies very close to the MPB. Peak intensities suggest that for 0.7BGF:0.3PT with Ga (x) = 0 the dominant phase is rhombohedral. However, with increasing x the tetragonal fraction appears to increase and the rhombohedral to decrease up to the stability limit $x \leq 0.1$ above which secondary, non-perovskite phases begin to appear. It is interesting to note the very large tetragonal distortion (separation of <001> and <100> peaks) at the MPB corresponding to a (c/a) ratio ~ 1.16 for 0.7 BGF:0.3PT with x = 0.1.[6]

For the tetragonal phase region, 0.4 BGF:0.6PT, a higher Ga (x) content was required to maintain high resistivity and the values of x in the range $0 < x < 0.5$ were explored. The <100> peak splitting shows that c/a first increases with Ga content up to x ~ 0.25, decreasing with further increments of x. However, for compositions above x = 0.25 there is evidence of secondary non-perovskite phases.

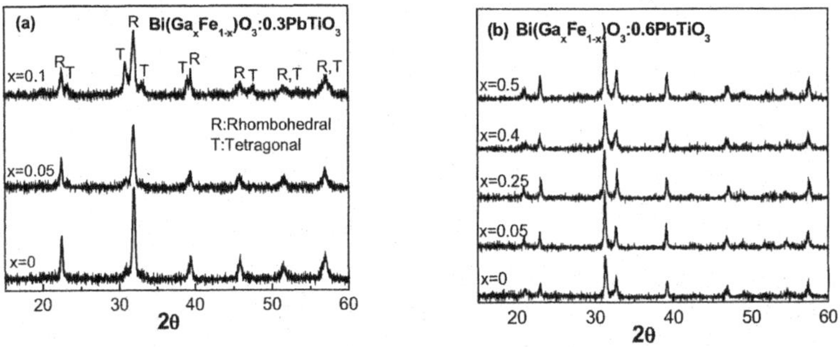

Fig. 8: XRD patterns of (a) BGF:0.3PT and (b) BGF:0.6PT for different Ga concentrations

For the lanthanum substituted composition BLGF:PT with just 5 mole % gallium substitution, phase pure perovskite structures with no secondary pyrocholore were observed for PT contents above 0.1 ($x \geq 0.1$). The structure changes from rhombohedral to tetragonal at the MPB. Figures 9 (a – c) show the lattice parameters as a function of PT content for various La concentrations, indicating how the rhombohedral to tetragonal transition moves to higher PT concentration with increasing La incorporation. It is very interesting to note the large overlap of rhombohedral and tetragonal phases around 0.3 PT in the La = 0 (x = 0) composition and the very large value of tetragonal c/a ~ 1.16 at this composition. High values persist up to La (x ~ 0.10) but drop markedly for x ~ 0.20 where c/a ~ 1.02. A tentative family of phase diagrams for the location of MPBs in the La modified family is given in Fig. 10 where the high temperature extreme were taken from dielectric measurements. At the lower La content the phase boundary is taken as the mid composition of the mixed phase region. At higher La concentration, as is evident in Fig. 9, the boundary is more clearly defined.

Perhaps the most important result of these studies is that depicted in Fig. 11 showing the manner in which electrical resistivity is improved in the 0.7 BGF:0.3PT and in the 0.4BGF:0.6PT.

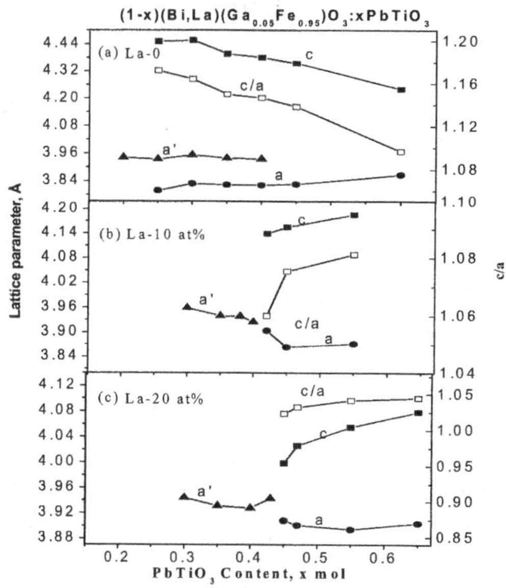

Fig. 9: Lattice parameters of BLGF:PT as a function of PT

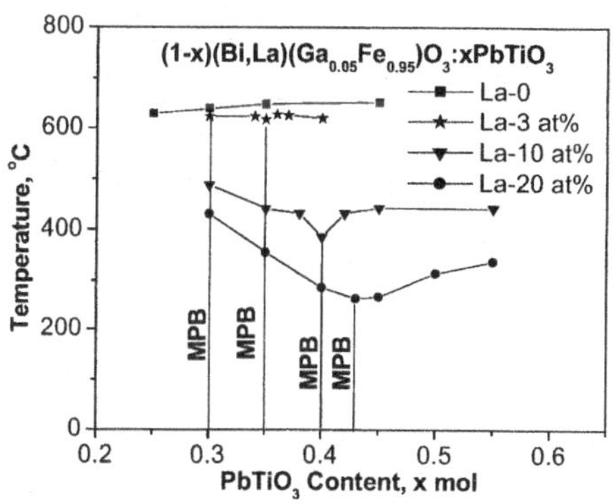

Fig. 10: Location of MPB in the BLGF:PT solid solutions for La concentration of 0, 3, 10 and 20 atom%

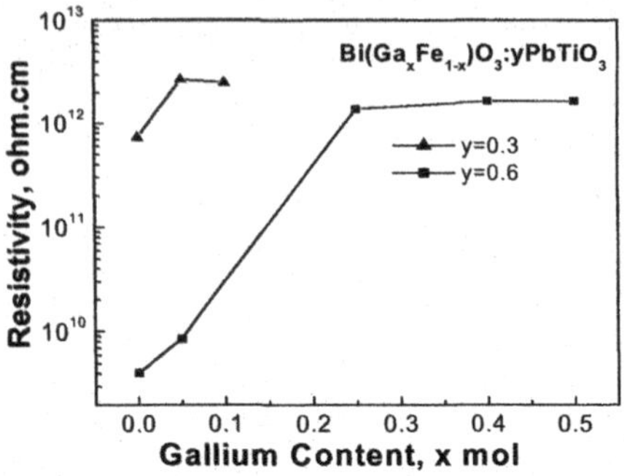

Fig. 11: Electrical resistivity of BGF:0.3PT and BGF:0.6PT as a function of Ga concentration

For the lanthanum substituted composition, it was possible to maintain an insulation resistance of over 10^{10} ohm cm with just 5 atom % gallium substitution with values of order 10^{13} ohm cm for the interesting range of the MPB compositions. This permitted excellent weak field dielectric permittivity measurements over the interesting range of PT as in shown in Fig. 12. As could be expected, the samples show weak dielectric dispersion, but clear peaks in K which track the composition of the MPB as determined by XRD.

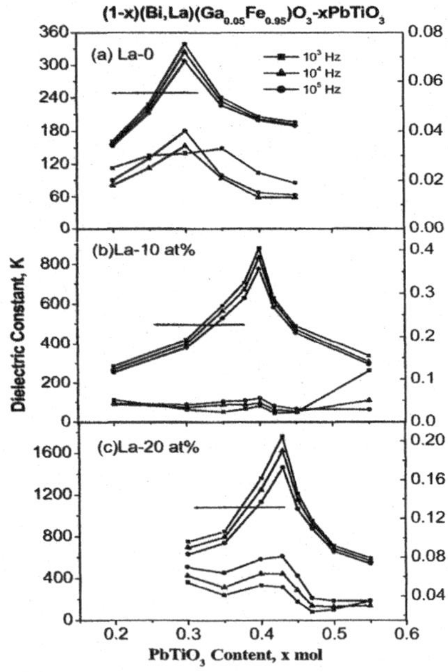

Fig. 12: Room temperature dielectric constant K and tanδ of BLGF:PT as a function of PT content for (a) 0; (b) 10; and (c) 20 atom % La.

High field dielectric hysteresis measurements are summarized in Fig. 13 taken on the modified Sawyer Tower circuit with a bipolar sinusoidal field at 10 Hz, up to 80kV/cm. Ferroelectric hysteresis is clearly evident in the samples with 10 and 20 atom % lanthanum with remanent polarization $P_r \sim 30\mu C/cm^2$, but E_c is still high at 20 kV/cm.

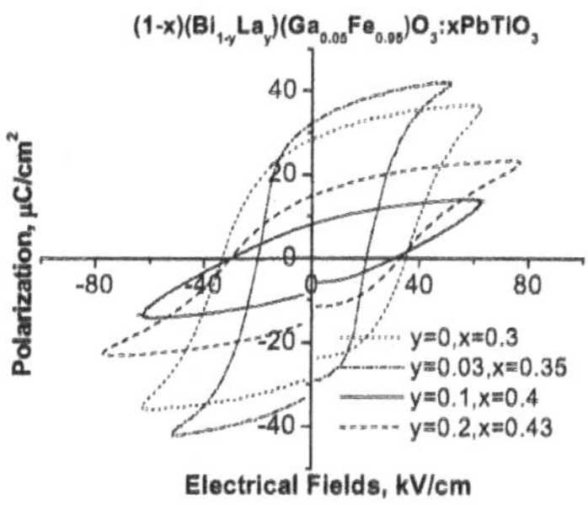

Fig.13: Dielectric hysteresis in BLGF:PT solid solutions taken with bipolar sinusoidal waves at 10 Hz

High field bipolar and unipolar strain S_{33} vs. E_3 are shown in Fig. 14. MPB compositions of BLGF-PT with 10 and 20 atom % of Lanthanum have similar strain levels of 0.35% and 0.2% under fields of 60 kV/cm. Estimating a large amplitude d_{33} value from a linear approximation to the slope gives $\Delta S_{33}/\Delta E_3 \sim 380$ pm/V. Measurement by Berlincourt meter and standard IEEE resonance method yield d_{33} values as a function of PT content for compositions with 10 and 20% La shown in Fig 15. Maximum values occur at the MPB compositions of 295 pC/N (20% La) and 163 pC/N (10% La) with planar coupling coefficients of order 0.36, confirming the response comparable to that of PZT ceramic.

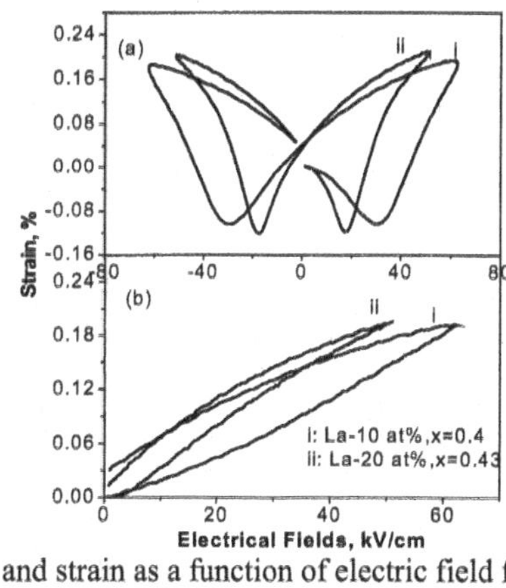

Fig. 14: Induced polarization and strain as a function of electric field for MPB compositions of BLGF:PT, (a) bipolar S –*E* response and (b) unipolar S –*E* response. Curves i and ii are corresponding to compositions of La-10 at%, x=0.4 and La-20 at%, x=0.43 respectively.

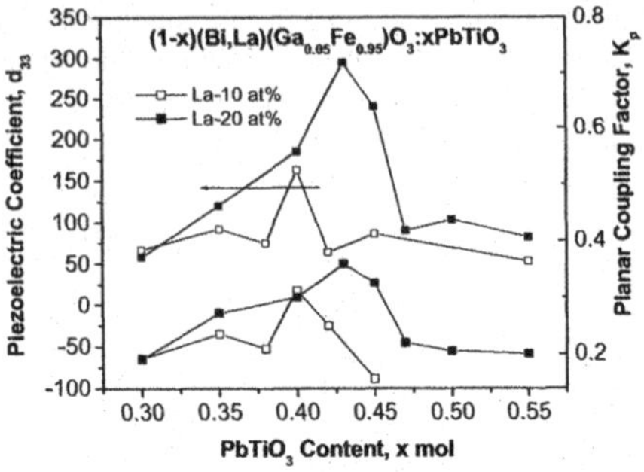

Fig. 15: Piezoelectric constant d_{33} and planar coupling factor K_p of BLGF:PT as a function of PT content with La concentrations of 10 and 20 atom %.

DISCUSSION

It is clear from the experimentation described that the BGF:PT solid solution can be fabricated as a single phase perovskite across the whole composition range and that it encompasses a ferroelectric rhombohedral to tetragonal phase transition. Small additions of gallium of order 5 atom % enhance the resistivity to ~ 10^{12} ohm cm and the currently processed ceramics will withstand very high electric fields up to 250 kV/cm. An unusual feature of the phase change in the BGF:PT system is the very broad composition range of co-existence of rhombohedral and tetragonal forms. A second unusual feature, which may be connected, is the truly massive c/a ratio of the tetragonal form of order of 1.16 at the MPB composition. In spite of the high breakdown field and the co-existence of rhombohedral and tetragonal forms it proved impossible to pole the 0.7 BGF:0.3PT composition for piezoelectric study.

Softening the composition as in PLZT by the addition of lanthanum, to form a BLGF:PT sequence, sacrifices some Curie Temperature but does sharpen the phase transition and at 10 atom % La the transition to cubic still occurs at 400°C and the c/a ratio of the tetragonal form is of order 6%. The purpose of the studies presented were to scope out some of the bounds and possible flexibilities in the properties which could be generated for compositions with morphotropic phase boundaries in the bismuth based perovskite family. Obviously now electrical conductivity is not a 'show stopping' problem, and strong piezoceramics with performance comparable to PZT and lower lead content can be generated. Clearly, the next step, as has now been accomplished for the bismuth scandate:lead titanate MPB composition, is to explore the growth of single crystals. With the very high Curie temperature of the ferrate, the parameter space available in the BLGF:PT system is much larger than in $BiScO_3$:PT compositions and the fact that very high tetragonal strain is preserved even in heavily La modified compositions gives confidence that even higher switchable strain may be possible.

An initial concern that the high coercivity persisting even in the La modified system might preclude the rotational instability which is essential for the high [001] activity in the crystals has been alleviated by the scandate studies, where even though the coercivity for ceramics switching is 20 kV/cm, as in BLGF:PT at the MPB,d_{33} ~ 1,250 pC/n and k_{33} ~ 90 % are shown to be achieved in <001> poled single crystals.[4]

Very recent studies in cooperation with Drs. Viehland and Dong in Virginia Tech show that the La modified BLGF:PT have substantial magneto-electric properties which will be the subject of further studies.

REFERENCES

[1]J. Cheng, W. Zhu, N. Li, and L. E. Cross, "Fabrication and Characterization of x$BiGaO_3$-(1-x)$PbTiO_3$: A High Temperature Reduced Pb-content Piezoelectric Ceramic", *Mater. Lett.* **57,** 2090 (2003).

[2]C.A. Randall, R.E. Eitel, T. R. Shrout, D. I. Woodward, and I. M. Reaney, "Transmission electron microscopy investigation of the high temperature $BiScO_3$–$PbTiO_3$ piezoelectric ceramic system", *J. Appl. Phys.* **93,** (11) 9271 (2003).

[3]Y. Shimojo, R. Wang, T. Sekiya, T. Nakamura, and L.E. Cross, "MPB Phase Diagram and Ferroelectric Properties in the $PbTiO_3$ System" Ferroelectrics, Vol **284,**121-128 (2003)

[4]S. Zhang, C. A. Randall, and T. R. Shrout, "High Curie temperature piezocrystals in the $BiScO_3$-$PbTiO_3$ perovskite system" *Appl. Phys. Lett.* **83** (15), 3150 (2003).

[5]Jinrong Cheng, Richard Eiel, Nan Li, and L.E. Cross, "Structural and electrical properties of $(1-x)Bi(Ga_{1/4}Sc_{3/4})O_3 - xPbTiO_3$ piezoelectric ceramics" *Journal of Applied Physics,* **Vol** 94 (1), 605-609 (2003)

[6]G. Smolenskii, B. Isupov, A. Agranovskaya, and N. Krainik, "New Ferroelectrics of Complex Composition", *Sov. Phys. Solid State* **2,** 2651 (1961).

[7]P. Fisher, M. Polomskya, I. Sosnowska, and M. Szymanski, "Temperature dependence of the Crystal and Magnetic Structures of $BiFeO_3$", *J. Phys. C.* **13,** 1931, (1980)

[8]Jinrong Cheng, L.E. Cross, "Effects of La Substituent on Ferroelectric rhombohedral/ tetragonal morphotropic phase boundary in $(1-x)(Bi,La)(Ga_{0.05}Fe_{0.95})O_3 - xPbTiO_3$ piezoelectric ceramics" Journal of Applied Physics, **Vol** 94 (8) 5188-5192, (2003)

THE WAGON WHEEL TRANSDUCER AS A VECTOR SENSOR AND A DIRECTIONAL PROJECTOR

A.-C. Hladky-Hennion
I.E.M.N., UMR 8520 CNRS,
41, Boulevard Vauban
59046 Lille CEDEX, France

J. K. Cochran Jr.
Georgia Institute of Technology
771 Ferst Drive, N.W.
Atlanta, GA 30332-0245, USA

R. Meyer and D.C. Markley
The Applied Research Laboratory
The Pennsylvania State University
University Park, PA 16802, USA

R. E. Newnham
The Materials Research Laboratory
The Pennsylvania State University
University Park, PA 16802, USA

ABSTRACT

Traditional flextensional transducers have been used extensively for fish-finding, oil exploration, sonar systems, and low frequency biomedical applications in the 1-100 kHz range. Our goal is to develop a family of small, low-cost, underwater and biomedical transducers based on extrusion technology. Some of these designs resemble traditional Class IV and Class VII flextensional transducers and will expand their useful operating range. Others incorporate features of tonpilz and vector sensor transducers. Miniature versions of these high-power, low frequency transducers are being optimized to produce broadband transmit and receive response, engineered vibration modes, and optimized acoustic beam patterns. Transducer topologies with various shapes, cross-sections, and symmetries can be fabricated through high volume, low-cost ceramic and metal extrusion processes. Dimensions of the pieces can be as small as one millimeter and as large as 10 cm, which provide resonance frequencies from above 1 MHz to below 10 kHz. The acoustic vibration modes, resonance frequencies and radiation patterns of these transducers are controlled through the symmetry and external shape of the transducer material, poling pattern, the driving and receiving electrode geometries and driving conditions. This is the first in a series of papers devoted to small, multimode flextensional transducers with *active* shells. The specific aim of this paper is to present a flextensional transducer shaped like a wagon wheel with active shells, which combines the advantages of small size and low-cost manufacturing with control of the shape of the acoustic radiation/receive pattern. The performance of the transducer as a vector sensor and as a directional projector is emphasized.

INTRODUCTION

As shown in Figure 1, traditional flextensional transducers generally consist of a piezoelectric transducer (the electromechanical driving element) and a vibrating shell which emits and receives acoustic waves. The shell is not piezoelectric. This is the first in a series of papers devoted to small, multimode flextensional transducers with active shells. There are several reasons why multimode transducers are of great interest.

First, a number of complex nonlinear phenomena are being studied for use in harmonic imaging, parametric arrays, elastography and sonoelasticity, and time-reversed acoustics. In these applications the transducer may be called upon to transmit at one frequency and receive at another, or to simultaneously transmit at two frequencies and receive at a sum or difference frequency, or to interrogate at one frequency and deliver focused high power at another, or to interrogate at high frequency while being continually repositioned at a much lower frequency, or to transmit in narrow beam patterns while receiving in an omnidirectional mode. Robust designs are needed for transducers that carry out two or more disparate functions, often at different frequencies, with different amplitudes and different phases.

A second important goal is the improvement of the bandwidth of the transducer. Many medical and underwater ultrasound transducers utilize small inexpensive broadband transmitters and receivers that function well in the low frequency range for long-range penetration and at higher frequencies for improved resolution at short range [1-3]. Transducers with these properties are normally obtained using an array of narrow-band transmitters, each operating at a different frequency, or by mass loading a narrow-band transducer. Combining the functions of two or more narrow-band transducers in one monolithic structure provides the benefits of more imaging information at a lower cost.

In addition to bandwidth, the shape of the acoustic radiation pattern and directional response of the transducers are also important parameters in ultrasonic applications. When the size of the transducer is small compared to the acoustic wavelength of interest, the transducer behaves almost as an acoustic monopole with omnidirectional response. However, this omnidirectional characteristic may be undesirable when the transducer is used as a projector, or in an array that is intended to radiate in only one direction. Often, large baffles are used, or rows of transducers are spaced and phased carefully to give the desired beam patterns. This can be an expensive, time consuming and cumbersome solution. A method has been developed for radially poled hollow cylindrical and spherical transducers with segmented electrodes driven together or individually in order to: (i) transmit and receive in certain directions by driving these sections with phase differences,

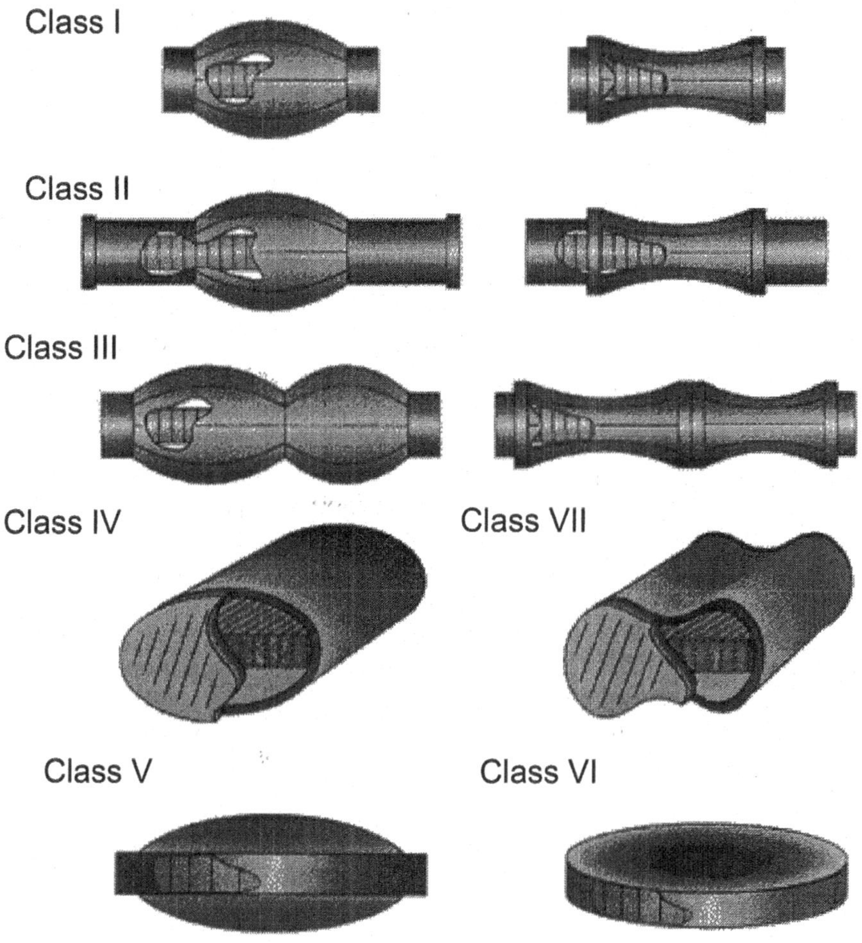

Figure 1. Standard Flextensional Transducers. Each of the designs consists of a metal shell driven by a stack of poled piezoelectric ceramics. In our method, we replace both the shell and the stack with a monolithic poled ceramic capable of generating multiple vibrations.

(ii) engineer the shape of the acoustic radiation beam patterns, (iii) identify the direction of incoming acoustic waves [4-6]. This idea was most recently applied to miniature cymbal flextensional transducers to obtain unidirectional transducers with cardioid beam patterns [7].

Miniaturization is also a general trend in biomedical and underwater transducers, but this poses certain performance problems. If the radiating dimensions of a projector are less than a wavelength ($ka \ll 1$), a large displacement is required to generate sufficient volume velocity to achieve high source levels. Since the typical strain that can be obtained from a bulk piezoelectric ceramic is limited to 0.1 %, amplification mechanisms are required to obtain the necessary large displacements for low frequency projectors. Flextensional transducers with flexible, compliant shells provide such a mechanism with the required volume displacements and 70 to 90 % efficiencies; however, the shells must accommodate both the hydrostatic stresses and the AC drive stresses [8]. Such mechanical amplification schemes add complexity to the design, cost to the fabrication, and raise reliability issues in the long-term use of the transducers.

The aim of this paper is to present a flextensional transducer with active shells, a transducer shaped like a wagon-wheel that combines the advantages of small size with control of the shape of the acoustic radiation pattern. In the first part of the paper, the transducer's design is described. Then, the fabrication and testing procedures are presented. Finally, the performance of the transducer as a vector sensor and as a directional projector is emphasized.

WAGON-WHEEL TRANSDUCER DESIGN

The wagon-wheel transducer (Figure 2) consists of a cylindrical shell with a small concentric tube at the center. The small tube is inactive, and intended mainly as a means of mounting a stress bolt or carrying a cable for power and driving circuits. Four arms span between the outer and inner cylinders. The outer shell and the arms can be activated simultaneously, or independently. The outer shell can be poled and driven from inside to outside or vice versa. The number of arms can also be increased to give the transducer a more robust design with possibly higher pressure tolerance at the expense of sensitivity, or decreased to provide a more compliant structure.

Complementing active arms with active shells provides several vibration modes, such as the radial mode of the shells, transverse mode of the arms, length modes of the entire structure, thickness mode of the walls, and various cross-coupled modes. All of these modes can be controlled through the dimensions of the structure and the way they are poled, activated, and driven. The resonance frequencies of these modes can be brought closer together to obtain a multi-

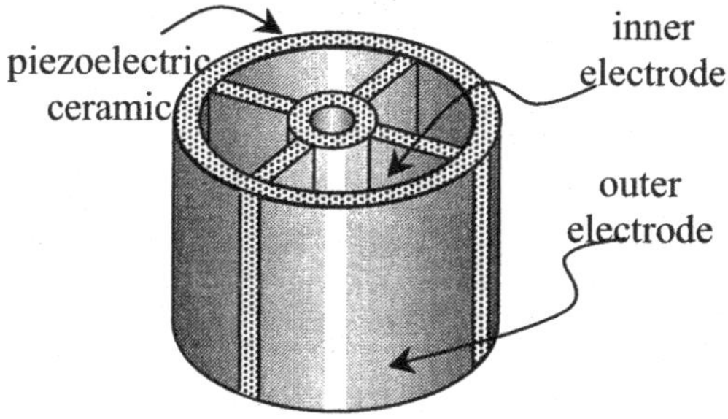

Figure 2. The Wagon Wheel, an example of a monolithic piezoceramic

frequency broadband transducer, or a transducer with dual operating frequencies, such as f and 2f. Figure 3 shows the many ways the segments can be poled and driven. The bold arrows indicate the electric field direction.

In addition to the geometrical and dimensional modifications, different sections of the transducer can be driven with a phase difference and with different voltages to control the acoustic beam pattern. A directional beam pattern can be achieved through cancellation of sound pressure in one direction and addition in the opposite direction. One method was initially developed by Erlich [4] for hollow cylindrical transducers with a plurality of electrodes to obtain dipole beam patterns. It was later applied to hollow spheres [5] and most recently to miniature flextensional cymbal transducers by Zhang [7] to obtain cardioid beam patterns.

TRANSDUCER FABRICATION AND TESTING

The monolithic ceramic transducer design presented in the previous section were formed by extrusion of lead zirconate titanate (PZT - $Pb(Zr,Ti)O_3$) piezoelectric ceramic compositions. The extrusion paste was prepared by dry mixing TRS610C powder (TRS Ceramics) with ethoxylated state acid (Lonza) lubricant (1 wt %) and A4M methylcellulose (Dow Chemical) binder (5 wt. %) in a high-speed blender for five minutes. (All weights are on a dry powder basis.) Water (18 wt %) was injected with a syringe into the blender while the powder was being mixed. This produced a finely granulated free flowing powder with a homogeneous moisture distribution. After waiting two hours for hydration of the binder, the granulated powder was compounded in a high shear Buss kneader. This produced a paste with a shear yield stress in the range of 400-500 KPa as measured with a capillary rheometer using the Benbow and Bridgewater technique. Paste rheological properties are controlled by the powder particle size/distribution, maximum solids content (ϕ_{max}), and volume fraction in the batch. Also, the type and volume of lubricant, type and volume of binder,

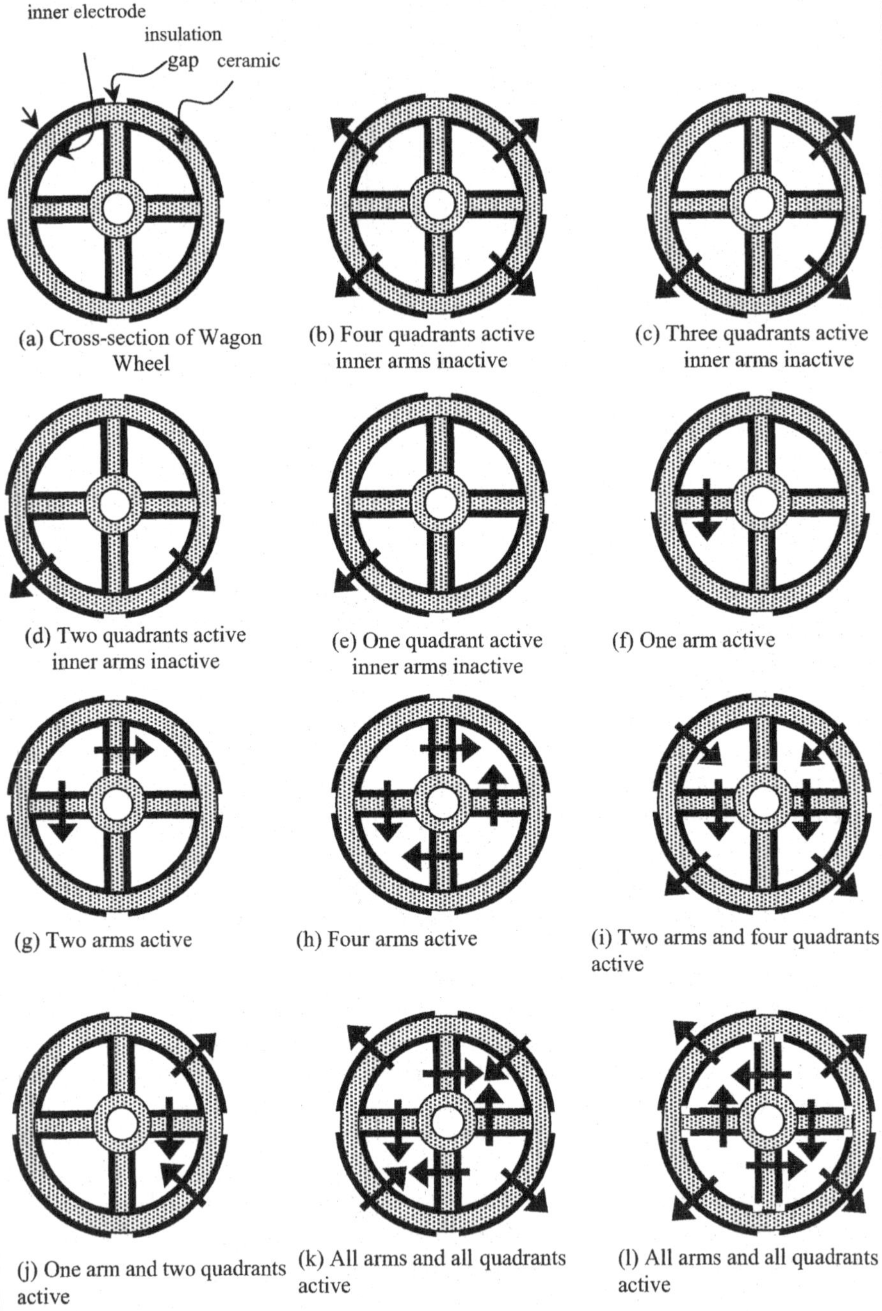

(a) Cross-section of Wagon Wheel

(b) Four quadrants active inner arms inactive

(c) Three quadrants active inner arms inactive

(d) Two quadrants active inner arms inactive

(e) One quadrant active inner arms inactive

(f) One arm active

(g) Two arms active

(h) Four arms active

(i) Two arms and four quadrants active

(j) One arm and two quadrants active

(k) All arms and all quadrants active

(l) All arms and all quadrants active

Figure 3. Possible driving conditions for the Wagon Wheel transducer design

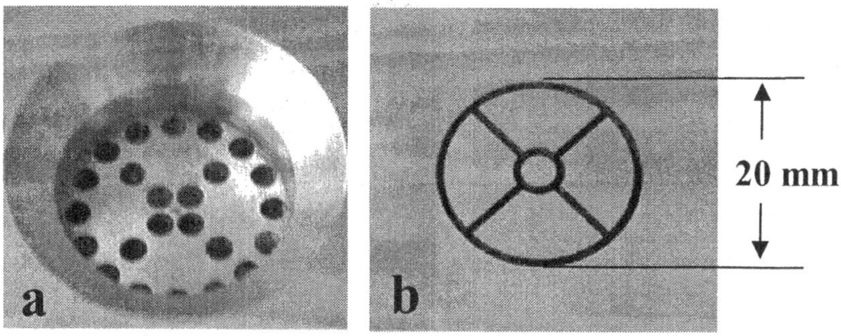

Figure 4. Linear wagon wheel extrusion die showing (a) feed holes for die entry and (b) die land formed by vertical EDM to produce a single piece die.

Figure 5. The extruded wagon-wheel transducer made from piezoelectric PZT with silver electrodes. It is intended for use as a directional receiver or emitter.

dispersants, and volume of water are rheology-controlling parameters. Control and prediction of paste properties have been reported elsewhere [9-12]. To produce a thin walled geometry with good precision, fine-grained powder with particle sizes about 1-2% of the feature dimensions are necessary. By using fine powders in the micron and submicron range, extruded articles with 100 μm features are feasible.

The wagon wheel transducer was formed with a 5 cm diameter piston extruder by forcing the paste through a linear single piece die with a wagon wheel die land, (Figure 4). Entry to the die was via 2.5 mm feed holes positioned with uniform spacing along the die land webs. The die land was 20.0 mm OD, 3.6 mm ID, and all webs were 0.80 mm thick.

In an initial experiment, a wagon wheel design with an average outer tube diameter (ϕ_o) of 17.5 mm, inner tube diameter (ϕ_i) of 5.0 mm and wall thickness (t) of 0.8 mm was fabricated from soft NAVY type VI material by extrusion (Figure 5). Ceramic pieces with various lengths were cut from the green piece and fired for binder burnout and sintering (550°C for 60 minutes and 1285°C for 90 minutes) under PbO rich atmosphere in a closed crucible. The dimensions after the sintering shrinkage are as follows: $\phi_o \approx 14.8$ mm, $\phi_i \approx 4.0$ mm, $t \approx 0.65$ mm, length (l) $\approx$ 12.5 mm. The fired pieces were dip coated with air-dry silver electrode and poled through the wall thickness at 120°C for 20 minutes under an electric field of 20 kV/cm. In this initial test, only the outer shell was driven and the arms were inactive.

Following the processing and fabrication procedure laid out in the previous paragraph, the resonant vibration modes were determined experimentally using admittance spectroscopy (HP 4194). The admittance spectrum of a transducer provides information about the resonance and antiresonance frequencies, their associated coupling coefficients and quality factors.

The displacement shapes and symmetry of the vibration modes are identified using the ATILA® finite element analysis (FEA) code. ATILA is an FEA code developed for the modeling of sonar transducers. It can provide information concerning prestresses, and the behavior under hydrostatic pressure (static analysis), together with the resonant frequencies of various modes and their associated coupling factors (modal analysis), the in-air or in-water impedance and displacement field, the transmitting voltage response, the free field voltage sensitivity and the directivity beam patterns (harmonic analysis) [13]. A three dimensional model of the wagon-wheel structure was created using the dimensions obtained from the sintered piece. Taking advantage of the symmetry of the structure, only 1/16th of the transducer was meshed (half length and one

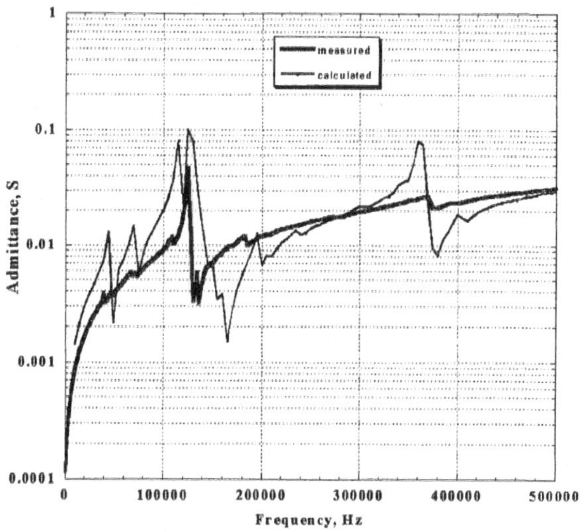

Figure 6. Measured vs. calculated admittance spectra of the wagon-wheel extrusion in air.

eighth of the section). ATILA mathematical solutions show excellent correlation with the measured results for the transducer in-air (Figure 6).

Figure 7 presents the displacement field of the wagon-wheel transducer at 44 kHz. It shows that, as the arm is expanding, the shell is retracting at $\pi/4$. At 70 kHz, the displacement is similar but both the arm and the shell at $\pi/4$ are expanding, whereas the shell at $\pi/8$ is retracting. At 70 kHz, a length motion is also observed. At 125 kHz, the frequency of the main peak on the admittance spectrum, the length mode is excited.

Next, the in-water performance of the wagon-wheel transducer can be modeled using the ATILA finite element code. Calculated outputs include impedance and admittance spectra, transmit and receive sensitivities, beam patterns, stresses, and displacements. This paper focuses on the receiving and transmitting performance of the device.

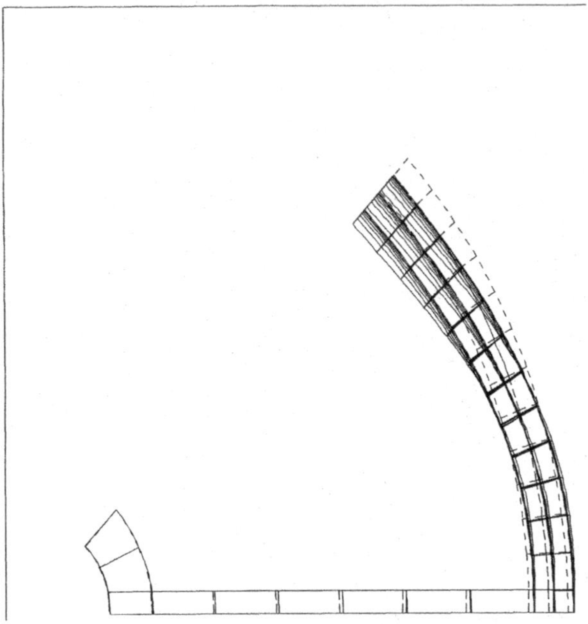

Figure 7. Displacement field of the wagon-wheel transducer at 45 kHz in a plane perpendicular to the length of the transducer. The dashed lines correspond to the rest position. Because of symmetry, only 1/16th of the transducer is meshed.

VECTOR SENSOR

In a number of undersea applications, it is desirable that the hydrophone possess directivity characteristics that reveal the direction of the source of incoming acoustic energy. Directional acoustic sensors have been in operation since 1969 as antisubmarine warfare platforms. Airborne (DIFAR) and activated sonobuoy (DICASS) systems both make use of pressure gradient sensors. The wagon wheel transducer is similar to the cylindrical designs used in activated sonobuoys. By monitoring the pressure-wave signals received on all four quadrants, both the direction and the gradient can be determined. Summing the four measurements gives the average pressure.

Operating principle

The wagon-wheel vector sensor operates as a quarter-wave phase detector. To visualize this mode of operation, consider a 2 cm diameter wagon-wheel

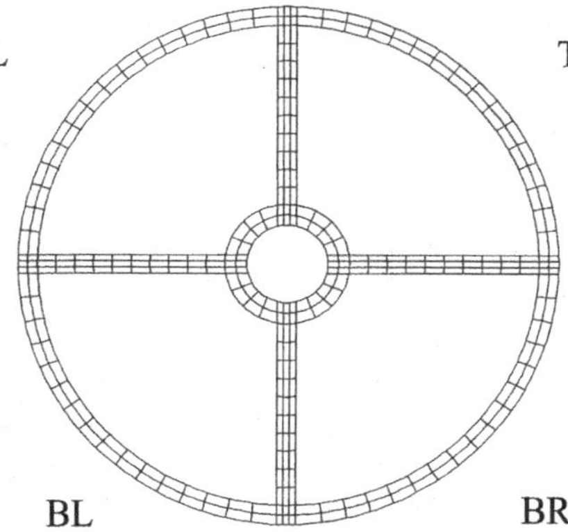

Figure 8: Bidimensional finite element mesh of the cross section of the wagon-wheel transducer.

operating in the 1-100 kHz range. At low frequency, down near 2 kHz, the wavelength in water is about 80 cm, 40 times larger than the transducer. At high frequency near 200 kHz, the wavelength is about 8 mm, significantly smaller than the 2 cm transducer.

When *d*, the diameter of the transducer, is much smaller than the wavelength, all four quadrants will measure the same amplitude and the same phase. In other words, the transducer acts as a point detector. When *d* is much larger than λ, the phase and the amplitude change rapidly within one quadrant and the difference between quadrants follows a complex function with small average values.

The phase differences between quadrants becomes much more apparent when $d = \lambda/4$ near 20 kHz. Here the front and back of the wagon wheel transducer are always out of phase, simplifying the mathematical description. The amplitude and phase change as the wave sweeps by the transducer leading to substantial phase differences of 90° to 180° between quadrants.

Next consider what happens when the incident wave arrives from different directions. Let θ be the orientation angle, as specified in Figure 8. Signals are detected in the four quadrants labeled top left (TL), top right (TR), bottom left (BL) and bottom right (BR). At $\theta = 0°$, the signals from TL and BL will be equal in magnitude and in phase. BR and TR are also equal in magnitude but 90° out of phase with TL and BL.

At $\theta = 45°$, BL and TR pass through maxima and minima, 180° out of phase. TL and BR pass through zero, 90° out of phase with TR and BL.

Numerical and experimental tests have been performed to evaluate the wagon-wheel transducer as a vector sensor. For the calculations, we assume that the length of the transducer is very large. In this way, a plane strain condition is obtained and the finite element mesh is reduced to a bidimensional mesh of the cross section in Figure 8. Only the shell is active, the arms are unpoled, and the transducer is immersed in water. A plane wave impinges on the transducer, at an angle θ with respect with the horizontal x axis. The angle θ is varied from 0° to 360° with the amplitude of the incident plane wave set at 1 Pa. Voltages are calculated for each of the four quadrants using ATILA finite element code (Figure 9a).

Figure 9b shows the results of testing a fabricated sample in water at 20 kHz. In spite of some asymmetry in the sample (one side was slightly flattened in green-handling), the measured results demonstrate that the predicted pattern is valid and attainable. With a perfectly symmetric sample an even better match should be easily obtained. Combining the voltages on the four quadrants and solving for the wave orientation angle θ gives:

$$\theta = Arctg\left(\frac{-V_{TR} + V_{BR} - V_{TL} + V_{BL}}{-V_{TR} - V_{BR} + V_{TL} + V_{BL}}\right) \quad (1)$$

Figure 10 shows the actual angle of the incoming plane wave versus that calculated from the measured voltages using equation (1). The agreement between the given angle and the deduced angle is very good. If a different frequency range is desired, the device parameters can be tailored to match.

Application

In sonobuoy deployment, several vector sensors operating at different frequencies are suspended from a flotation platform along with an electronics package and battery. A cable spool and heavy mass are used to control the depth and verticality of the sonobuoy. Wagon-wheel vector sensors 1-4 cm in diameter would cover the 10-40 kHz range. An antenna positioned within the flotation sphere relays the signals to neighboring stations. Harmonic imaging in the 20-80 kHz range could be carried out using the fundamental as a pulsed directional transmitter and second harmonic as a vector receiver, similar to the process used in biomedical ultrasonics.

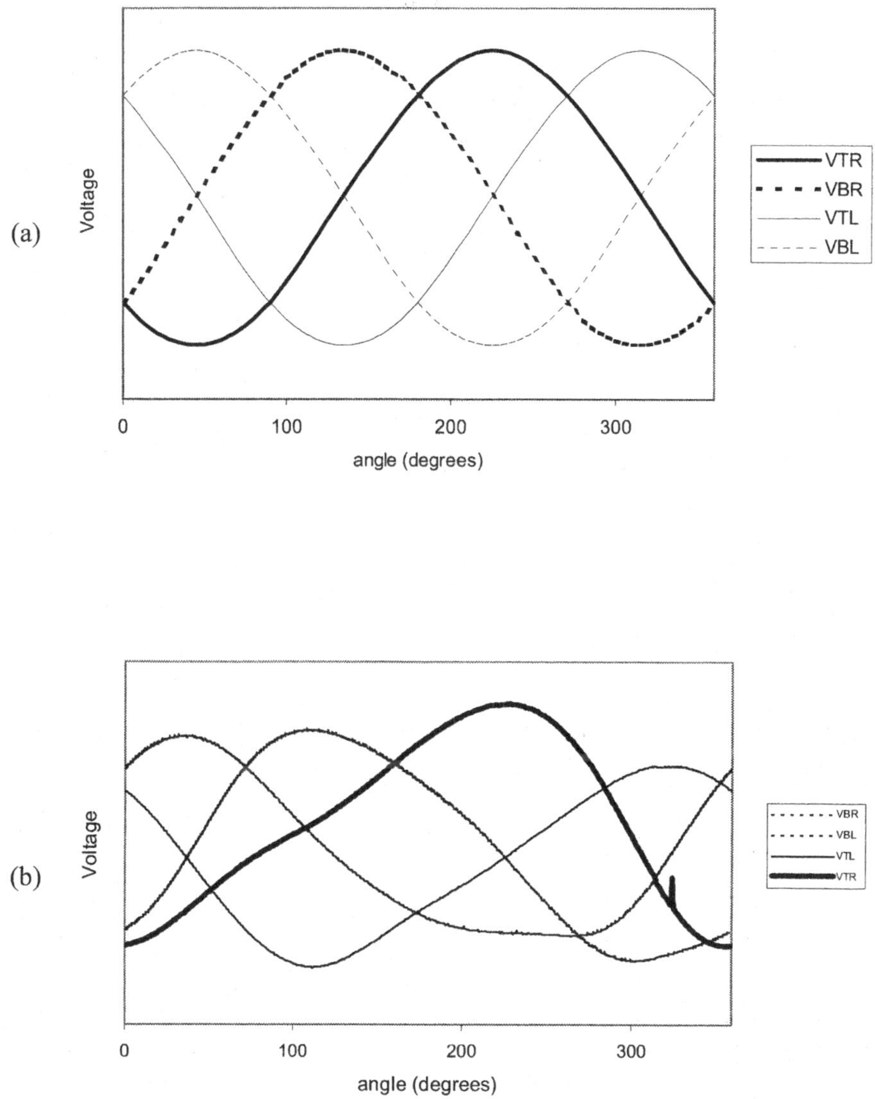

Figure 9: Numerical (a) and experimental (b) results: variations of the modulus of the four voltages as a function of θ, the angle of the incident plane wave of 1 Pa impinging the wagon-wheel transducer at 20 kHz.

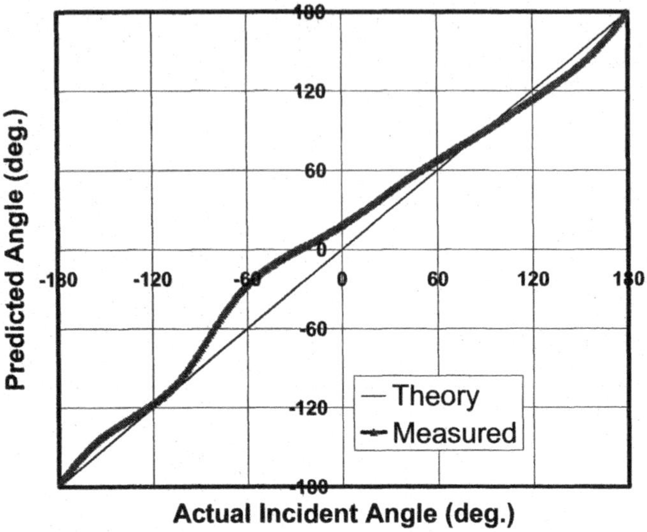

Figure 10: Chart showing predicted vs. experimentally determined incoming plane-wave angle.

VECTOR PROJECTOR

In this section, first the performance of the wagon-wheel transducer as an underwater projector is discussed, and a technique to evaluate the beam pattern is presented.

The wagon-wheel projector

Finite element calculations were performed on the wagon-wheel transducer to give the Transmitting Voltage Response versus frequency. For the sake of simplicity, the length of the transducer is assumed to be very large, thus a plane strain condition is used, and the finite element mesh is reduced to a bidimensional mesh. Three different excitation cases were considered: first, only the shell is active, the arms are unpoled and one Volt is applied on each quadrant; second, only the arms are active, the shell is unpoled and one Volt is applied across the thickness of each arm; finally, both arms and shell are active. These three configurations are compared with an empty cylinder of the same diameter as the wagon-wheel transducer.

When the shell is active, the wagon-wheel transducer gives a TVR of same order of magnitude than the cylinder. At low frequencies (<40 kHz), the wagon-wheel transducer has a higher TVR than the cylinder but the bandwidth is narrower.

When only the arms are active in the wagon-wheel transducer, the efficiency is low. One can adjust the phase of the voltage in the arms or the poling direction in the arms, with a view to combining the output of the active shell and of the active arms.

General principle of the multiple-drive transducer

With four leads attached to the inner surfaces, each quadrant can be driven independently by adjusting the phase, frequency, and amplitude of the applied voltage. Combining these possibilities in the optimum manner provides a useful method of controlling the acoustic beam pattern. The procedure is similar to that carried out earlier with the Double-Driver cymbal transducer [7]. An advantage of the wagon-wheel is that beam patterns can be controlled in two-dimensions. The earlier work was mainly one-dimensional. One application envisaged for the transducer is in the unmanned underwater vehicles and automated sonobuoys needed for harbor and fleet security.

For sake of simplicity, the principles of the multiple-drive operation are illustrated in Figure 11 showing four different drive possibilities. The goal is to generate a cardioid beam pattern. In this case, only the shell is active, leaving the arms unpoled. Because of the geometry of the wagon wheel transducer, even more complex beam patterns can be obtained. If the inside electrode is used as the ground electrode, and the four outside electrodes are driven in phase with the same electric field ($V_{TR} = V_{TL} = V_{BR} = V_{BL}$), as shown in Figure 11.1, an omnidirectional beam pattern (monopole) will be obtained (Figure 11.2). If the two outside electrodes of the top part (V_{TR} , V_{TL}) and the two outside electrodes of the bottom part (V_{BR}, V_{BL}) are driven under electric fields with the same amplitude but a phase difference of 180°, then a dipolar beam pattern will be obtained (Figure 11.3). In the dipole mode, the Transmit Voltage Response (TVR) shows maxima in the front and back of the transducer, but their phase is opposite. If the output from the dipole mode is added to the output of an omnidirectional mode with equal TVR, a cardioid beam pattern with only one maximum can be obtained (Figure 11.4).

The procedure defined above is for the simplest case where only two pairs of electrodes are involved in the structure. By switching the coupled electrodes (i.e. V_{TR} and V_{BR} as front and V_{TL} and V_{BL} as the back electrodes and so on) the cardioid beam pattern can be rotated around the circumference of the cylinder. Incorporating the double drive idea into even more complex designs with active arms provides an enormous number of ways to shape the acoustic radiation patterns.

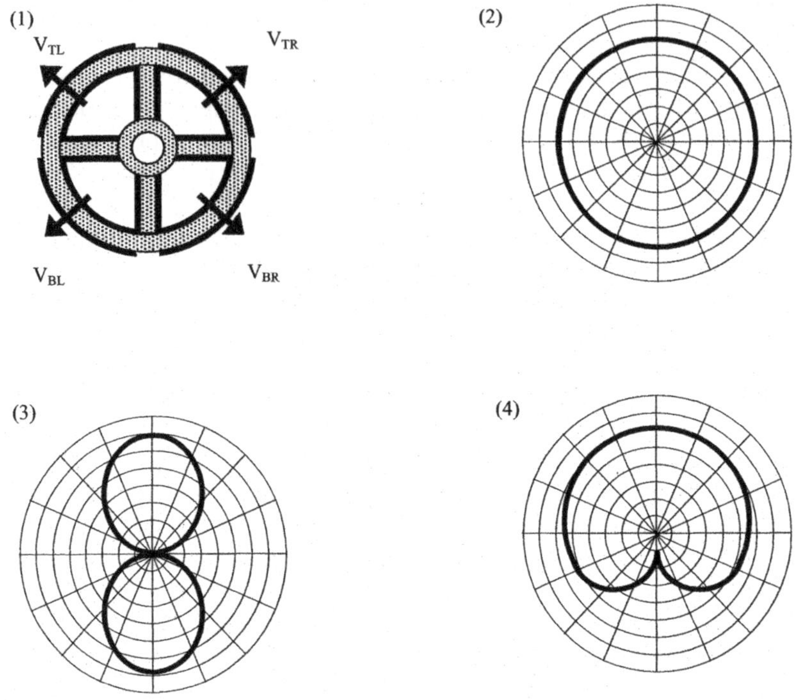

Figure 11. The driving scheme of the wagon-wheel transducer and resultant beam patterns.

(1). Cross-section of the transducer with separated outer electrodes and common ground inside electrode. Each quadrant is poled is the thickness. Arms are unpoled. Four different voltages can be applied.
(2). Monopole– omnidirectional mode ($V_{TR} = V_{TL} = V_{BR} = V_{BL}$). Level is TVR_m
(3). Dipole mode ($V_{TR} = V_{TL} = - V_{BR} = - V_{BL}$). Level is $TVR_d \sin(\theta)$
(4). Cardioid mode ($V_{TR} = V_{TL} == V_f$, $V_{BR} = V_{BL} = V_b$, with $\left(\frac{V_b}{V_f} = \frac{(1-R)}{(1+R)}\right)$, where

$R = \frac{TVR_m}{TVR_d}$. Level is $TVRc(1+\sin\theta)$

Mathematically the cancellation procedure goes as follows [7]:

Let V_f be the electric field applied to the front side and V_b the electric field applied to the back. V_m and V_d are driving fields associated with the monopole and dipole drive conditions.

$$V_f = V_m + V_d \tag{2}$$

$$V_b = V_m - V_d \tag{3}$$

$$\frac{V_b}{V_f} = \frac{1-r}{1+r} \quad \text{where } r = \frac{V_d}{V_m} \tag{4}$$

TVR, the Transmit Voltage Response, is related to the voltage by

$$TVR_b = \frac{p_b}{V_b} \quad \text{and} \quad TVR_f = \frac{p_f}{V_f} \tag{5,6}$$

where p is the measured sound pressure. To cancel the sound pressure completely in the back side, the pressure amplitudes should be equal, leading to :

$$\frac{V_b}{V_f} = \frac{1-R}{1+R} \quad \text{where } R = \frac{TVR_m}{TVR_d} \tag{7}$$

The complex ratio R is determined from the measured monopole and dipole TVRs. The equation gives the ratio of the voltages and the phase lag φ on each side of the transducer. The values of the applied voltage are complex.

Examples of directional beam patterns obtained by the wagon –wheel transducer

For simplicity, first only the shell is active, and the arms are unpoled. The frequency is once again chosen equal to 20 kHz, as explained previously. With a view to doing comparison, the TVR for the wagon-wheel transducer with active shell is 156.22 dB, and the other configurations give very similar levels. In this case, a directivity function of the following form is obtained:

$$A+B\cos\theta+C\sin\theta+D\sin 2\theta \tag{8}$$

In this expression the A, B, C and D coefficients are selected to give the desired beam pattern. One might, for example, maximize the beam in a given direction and minimize it in other directions.

(a) (b) (c)

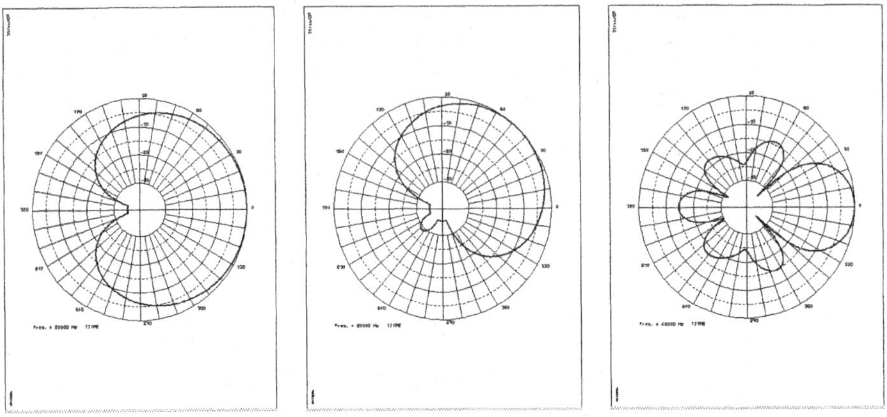

Figure 12. Directional beam pattern obtained with the wagon-wheel transducer. (a) Cardioid at 20 kHz, (b) directional beam pattern at 45° at 20 kHz, (c) more complex directional beam pattern at 0° at 40 kHz.

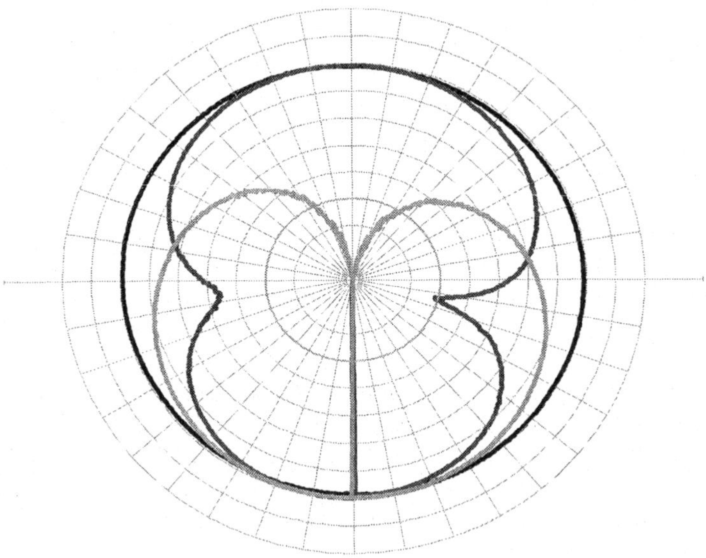

Figure 13. Experimental monopolar, dipolar and cardioid beam patterns obtained with the wagon-wheel transducer at 20 kHz.

The classical directional beam pattern is a cardioid (Figure 12.a). Following the previous principle, the voltages applied on each quadrant are: $V_{TR} = V_{BR} = 1$ and $V_{TL} = V_{BL} = -0.94 + 0.46\,j$. The resulting TVR at 0° is 149.27 dB, 143.19 dB at 90°, and 105.7 dB at 180°. Compared with the previous value of 156 dB, the TVR is significantly reduced in the cardioid case.

Next consider the following directivity function: $1+\cos\theta+\sin\theta+\frac{1}{2}\sin 2\theta$. This directivity has a maximum at $\pi/4$, and is equal to zero at π and $3\pi/2$. The level at $5\pi/4$ is low. Following the previous principle, the voltages applied on each quadrant are : $V_{TR} = 1$, $V_{BR} = V_{TL} = -0.97+0.32j$, and $V_{BL} = 0.79-0.61j$. The corresponding directivity is presented on Figure 12.b. The TVR at $\pi/4$ is 136.72 dB and at 0° it is 133.52 dB. As expected, when the directivity is improved, the power level grows smaller.

More complex directivity patterns can be obtained when the arms are active. Under these circumstances the $\cos 2\theta$ and $\cos 4\theta$ functions become available at higher frequencies. Figure 12.c presents a more complex directivity pattern utilizing active shell and arms at 40 kHz. Once again, the directivity is improved because the main maximum is more than 15 dB higher than the other maxima, but the TVR is again reduced. Moreover, the applied voltages have to satisfy a more complex solution.

This technique has been experimentally tested on the previous sample to obtain a directional pattern for the wagon wheel transducer. Figure 13 presents the experimental directivity pattern at 20 kHz, showing clearly a monopolar, dipolar and a cardioid shape.

CONCLUSION

Compared to the standard flextensional designs, the multimode monolithics are much less expensive because no assembly is required and there are no glue bonds. They can be extruded to any length and to a wide range of diameter, giving an extensive range of resonant frequencies. But most important of all, carefully designed dies make multimode operation possible for wider bandwidth, beam shaping, and many interesting applications in nonlinear acoustics. Multiple drive operation of a wagon-wheel transducer made by extrusion is of interest for vector sensor applications and for directional projector applications.

ACKNOWLEDGEMENTS

The authors would like to acknowledge the support of Jan Lindberg (Code 321) at the Office of Naval Research (N00014-01-1-0872).

REFERENCES

1. S.C. Butler, and F.A. Tito, "A Broadband Hybrid Magnetostrictive/ Piezoelectric Transducer Array," *Proceedings of the MTS/IEEE Oceans 2000 Conference and Exhibition*, **3** 1469 –1475 (2000).

2. T.R. Howarth, and R.Y. Ting, "Development Of A Broadband Underwater Sound Projector," *Proceedings of the MTS/IEEE Oceans 1997 Conference and Exhibition*, **2** 1195-1201 (1997).

3. J. Lan, and S.G. Boucher, "Dual Frequency Mode Transducer" *Proceedings of the MTS/IEEE Oceans 1996 Conference and Exhibition,* **2** 763-766 (1996).

4. S.L. Erlich, "Sonar Transducer," U.S. Pat. No. 3,290,646, 1966

5. S.L. Erlich, "Spherical Acoustic Transducers," U.S. Pat. No. 3,732,535 1973.

6. S.H. Ko, G.A. Brigham and J.L. Butler, "Multimode Spherical Hydrophone," *Journal of the Acoustical Society of America*, **56**, [6] 1890-1898 (1974).

7. J. Zhang, A.-C. Hladky-Hennion, W.J. Hughes and R.E. Newnham, "A Miniature Class V Flextensional Cymbal Transducer With Directional Beam Patterns: The Double Driver," *Ultrasonics*, **39** 91-95 (2001).

8. E.F. Rynne, *Innovative Approaches For Generating High Power, Low Frequency Sound in Transducers for Sonics and Ultrasonics*; pp. 38-49 Technomic Publishing Co., Inc., Lancaster, PA, 1993.

9. K.M. Hurysz, R. Oh, J.K. Cochran, T.H. Sanders, and K. J. Lee, "Modeling Powder Extrusion Pastes for Forming Light Weight Structures", Processing/Properties of Lightweight Cellular Metals and Structures, Amit Gosh, Tom Sanders, and Dennis Claar, eds., pp 167-176, TMS, (2002).

10. K. M. Hurysz and J. K. Cochran, " The application of high solids content suspension models to pastes", Journal of the European Ceramic Society, **23** 2047-2052, (2003).

11. R. Oh, K. M. Hurysz, J. K. Cochran, "Effects of Die Geometry and Paste Rheology on Extrusion Pressure", Innovative Processes/Synthesis: Ceramics, Glasses, and Composites VII, N.P. Bansal and J. P. Singh, eds. Ceramic Transactions, A. Cer. S., In Press, (2003).

12. K. M. Hurysz, and J. K. Cochran, "Bulk Shear Stress Modeling of Pastes Made from Low Maximum Solids Content Raw Materials", Innovative Processes/Synthesis: Ceramics, Glasses, and Composites VII N.P. Bansal and J. P. Singh, eds., Ceramic Transactions, A. Cer. S., In Press, (2003).

13. A.-C. Hladky-Hennion, and J.N. Decarpigny, "Finite Element Modeling Of Active Periodic Structures: Application To 1-3 Piezocomposites," *Journal of the Acoustical Society of America*, **94** 621-635 (1993).

EFFECT OF CATHODE MATERIALS ON THE PERFORMANCE OF SINGLE CHAMBER SOLID OXIDE FUEL CELLS AND MODULE

Fatih Dogan, Toshio Suzuki, Piotr Jasinski and Harlan U. Anderson.

Department of Materials Science and Engineering and
Electronic Materials Applied Research Center
University of Missouri-Rolla
222 McNutt Hall, Rolla, MO 65401 USA

ABSTRACT

The effect of electrode materials on the cell voltage of single chamber solid oxide fuel cells (SC-SOFC) was investigated. SC-SOFC, operated in the mixture of fuel and oxidant gas, generated a cell voltage due to catalytic reactions of fuel on the electrodes. Typical cathode materials (perovskite oxides) are known as catalytically active for fuel oxidation. Cells with a Sm-doped ceria (SDC) electrolyte and a mixture of SDC and (La, Sr)(Co, Fe)O_3 (LSCF) as the cathode, and Ni-SDC as the anode were prepared and tested in single chamber configuration. The open circuit voltage decreased as the SDC content decreased in the LSCF cathode and dropped ~0.1V when pure LSCF was used as a cathode. The overpotential resistance also decreased with the decreasing SDC content in the LSCF. A power density of ~140 mW/cm^2 was obtained at 600°C using Ni-SDC anode and 10% SDC–LSCF cathode. A module, which consists of a SDC electrolyte with two Ni-SDC anodes on one side and two SDC-LSCF cathodes on the other side, has also been tested in SC-SOFC condition. It was shown that the two cell module generated 1.5V open circuit voltage with a maximum current of ~50 mA at 550°C.

INTRODUCTION

High temperature electronics require power sources to withstand the extreme temperature conditions. Solid oxide fuel cells (SOFC), operating at 800~1000°C are possible candidates as reliable power sources which have the potential to be utilized as integrated components in high temperature device applications. Recent studies on single chamber SOFC (SC- SOFC) lead to relatively high performance with distinct advantages as compared to double chamber fuel cells [1-3]. In SC-SOFC, both the anode and cathode are exposed to the same mixture of fuel and oxidant gas. Since SC-SOFC is operated using a mixture of fuel and oxidant gases in a simple cell structure without sealing materials so that a better thermal and mechanical stability, and direct use of hydrocarbon fuel can be achieved [3].

The oxygen activities at the anode and the cathode are not fixed and one of the electrodes has a higher electro-catalytic activity for the anodic oxidation of fuel, whereas the other electrode has a higher electro-catalytic activity for the cathodic reduction of oxygen. As a result, an EMF between two electrodes is generated even in the atmosphere of a fuel/air mixture. The following simplified oxidation and/or reforming reactions may occur on the electrodes using propane as fuel:

$$C_3H_8 + 3/2O_2 \rightarrow 3CO + 4H_2 \quad (1)$$
$$H_2 + O^{2-} \rightarrow H_2O + 2e^- \quad (2)$$
$$CO + O^{2-} \rightarrow CO_2 + 2e^- \quad (3)$$

The reaction 1 is exothermic evolving heat, which effectively increases the cell temperature. It is known that such reactions can take place not only on the anode but also on the cathode, since the typical cathode materials such as perovskites exhibit catalytic activity for hydrocarbon oxidation [4].

Our studies have been focused on several cathode and electrolyte materials in SC-SOFC to investigate their role on the performance of SC-SOFC [5-7]. This paper will address the development of SC-SOFC as a sealant-free fuel cell, which generates sufficiently high power densities for various high temperature electronic applications. The role of cathode composition on the on the open circuit voltage and the performance of SC-SOFC will be discussed.

EXPERIMENTAL PROCEDURE

Samarium doped ceria (SDC) electrolyte was prepared by pressing of commercially available powders (20 % samarium doped ceria, Daiichiki Genso Co, Japan) and sintering at 1500°C for 10 hours in air. After screen-printing of 70 wt%Ni-SDC ink as anode material on the electrolyte, the sample was sintered at 1450°C for 1 hour. Subsequently, mixtures of $La_{0.8}Sr_{0.2}Co_{0.2}Fe_{0.8}O_3$ (LSCF, SSI Technologies Inc., Janesville, WI) and SDC powder (0, 10, 30 wt% SDC) were screen printed on the opposite side of the electrolyte and sintered at 1000°C for 1 hour in air.

The electrochemical properties of the cells were measured in a tube furnace using a flowing mixed gas with a composition of 10 % propane and 90 % air as shown in Fig. 1. The furnace- and sample-temperatures were monitored simultaneously. Pt and Au mesh, in the size of electrode area, were used as current collectors. Furnace temperature was adjusted between 500 and 650°C. The flow rate of mixed gas was kept constant at 300 cm^3/min using a gas flow controller. To measure the open circuit voltage (OCV) of cathode materials, Ag was used as a reference electrode (Ag paste, Electro-Science Laboratories). Typical dimensions of the samples were 0.18 cm^2 electrode area and 0.5 mm electrolyte thickness. The impedance spectroscopy techniques were applied to investigate the cell performance using Solartron 1470 Battery Tester and 1255B Impedance Gain Phase Analyzer.

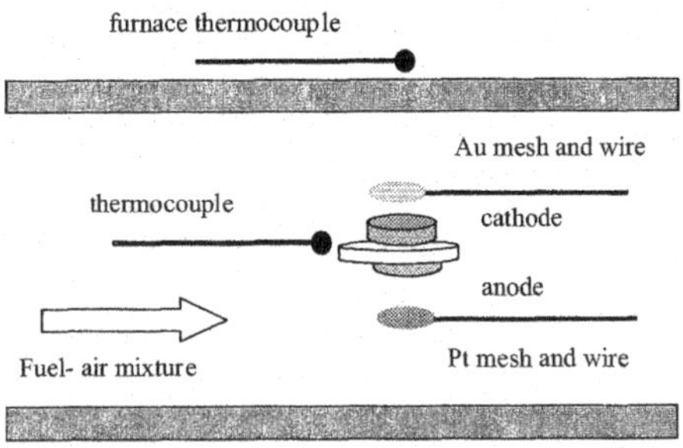

Figure 1: Experimental set up for SC-SOFC measurement

RESULTS AND DISCUSSION

Figure 2 shows SEM images of a cell with Sm-doped CeO_2 (SDC) as an electrolyte, Ni-SDC as an anode and LSCF as a cathode prepared using tape cast and screen printing processing techniques. Figure 3 shows the voltage between Ag electrode and the cathode materials as a function of SDC content in the LSCF. The negative sign of the voltage, measured as high as 0.1 V, is related to the LSCF cathode acting as an anode against the Ag electrode. This suggests that LSCF exhibits a catalytic activity promoting the hydrocarbon oxidation.

Figure 4 shows the open circuit voltage (OCV) of the cells with Ni-SDC anode as a function of SDC content in LSCF along with the results obtained using Ag electrode as cathode. The OCV, generated from Ni-SDC anode is as high as 0.9 V with the Ag electrode. It is shown that the OCV increased with increasing amount of SDC in LSCF. These results are consistent with those shown in Fig. 3.

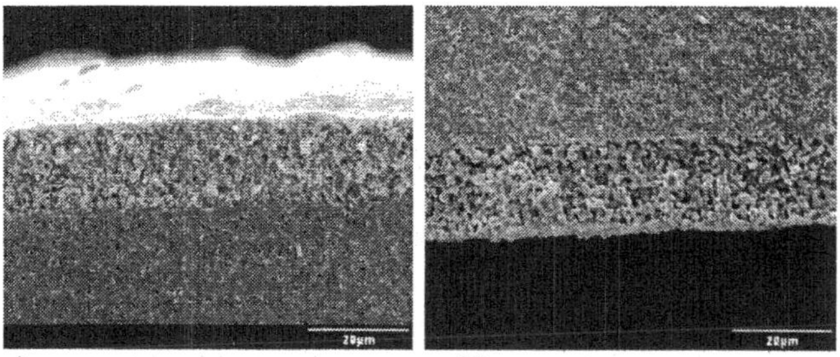

a) **b)**

Figure 2: Microstructure of electrolyte supported ceria based fuel cell; a) cathode / electrolyte interface, b) electrolyte / anode interface.

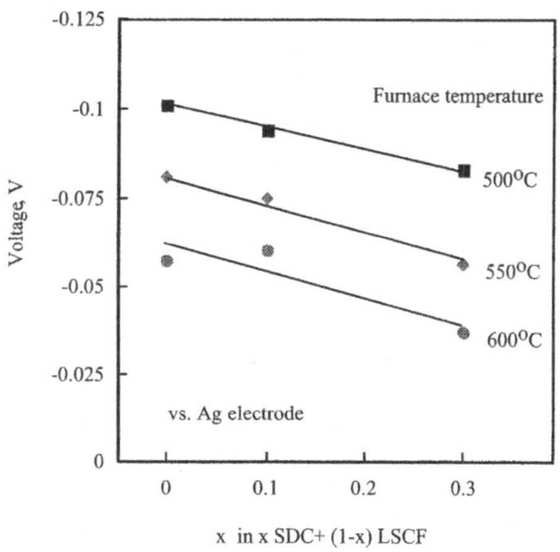

Figure 3: The voltage between cathodes and Ag electrode as a function of SDC content in the LSCF electrode.

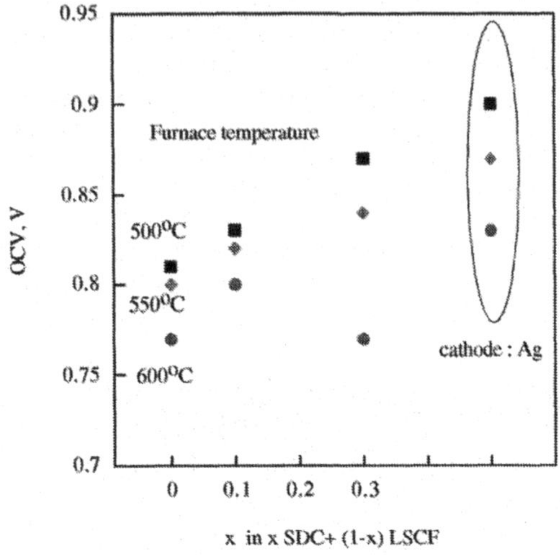

Figure 4: Open circuit voltage as a function of SDC content in the LSCF electrode with Ni-SDC anode.

The OCV decreased as the temperature and the SDC content of the cathode increased. An increase of SDC content decreases the catalytic activity of the electrode leading to a lower reaction rate for Eq. 1, which in turn, results in lower OCV. The decrease of OCV with increasing temperature may be related to the shift of main fuel-oxygen reaction from partial to full oxidation. This reaction may lead to a decrease of H_2 concentration at the electrode resulting in a decrease of the OCV. It may be confirmed by determining the catalytic activity of the electrode materials, which is presently under investigation. Note that the results shown in Figures. 3 and 4 can also be affected by the surface area, thickness and size of the electrode materials. Nevertheless, the use of LSCF as cathode material results in reduced OCV, which leads to lowering of the cell performance.

Figures 5 and 6 show the overpotential and electrolyte resistance of the cell as a function of SDC content in LSCF cathode. The overpotential and electrolyte resistances are determined from impedance spectra, where they correspond to the size of semi-circle and the high frequency intercepts of the semi-circle, respectively. As can be seen, the cell with the 30SDC–LSCF cathode showed the lowest overpotential resistance, which suggest that this composition leads to improved performance of SC-SOFC. The electrolyte resistance, however, resulted in lowest value when LSCF was used as cathode. This can be explained by the change of cell temperatures due to the different catalytic activities at the electrode caused by increasing amount of SDC in LSCF. This is consistent with the results shown in Fig. 3, so that a higher voltage (higher catalytic activity) was observed when SDC content in LSCF decreased.

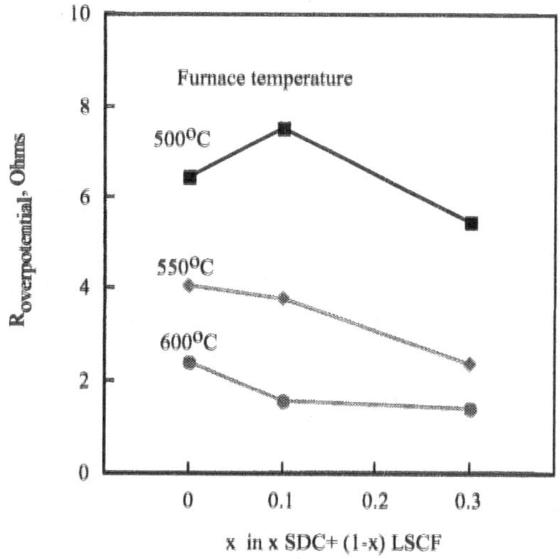

Figure 5: The overpotential resistance as a function of SDC content in LSCF cathode.

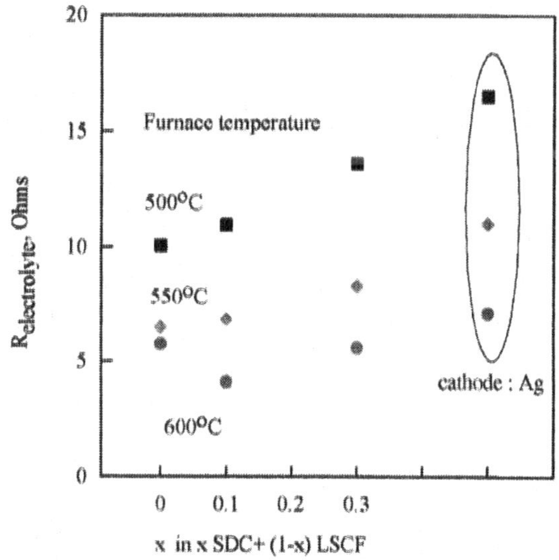

Figure 6: The electrolyte resistance as a function of SDC content in LSCF electrode

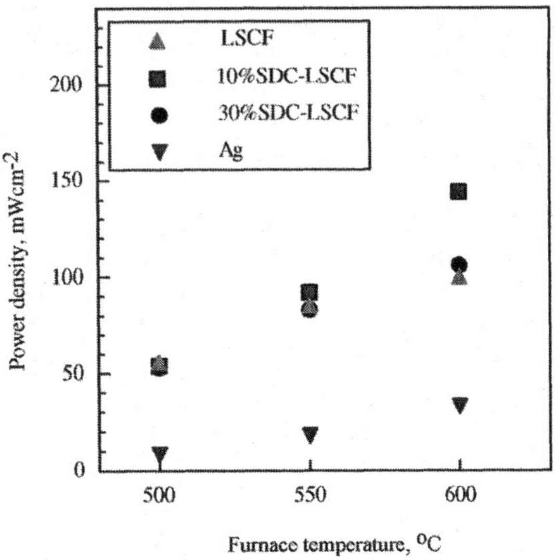

Figure 7: Maximum power density of the electrolyte supported fuel cells as a function of furnace temperature for different cathode in single chamber SOFC condition. Electrolyte thickness: 0.5 mm.

Figure 7 shows the maximum power density of SC-SOFC with different cathode materials at furnace temperatures from 500 to 600°C. Among tested samples, the cell with 10SDC-LSCF cathode achieved the highest power density of 140mW/cm^2 at 600°C. The lower performance of the cell with 30SDC-LSCF cathode may be due to higher electrolyte resistance, resulting from lower catalytic activity of the cathode or possible crack formation in the sample which was observed after measurement of the cell with 30SDC-LSCF cathode.

As discussed above, the cell performance cannot be evaluated only by the overpotential of the cathode, but also the catalytic effect of the cathode should be considered. Obtaining of higher values for OCV and improved cell performance will require thinner electrolytes as well as the development of cathodes and anodes with lower and higher catalytic activity for fuel oxidation, respectively.

SC-SOFC Module

Figure 8 shows a schematic diagram of SC-SOFC module. A module with two cells was prepared on SDC electrolyte (5 x 10 mm) using the same technique as described previously using Ni-SDC anode and 10 SDC-LSCF cathode with the area of 3 x 3 mm for each cell. The cells on the SDC were connected with gold line painted on the SDC using a gold paste (Fig. 8). Pt and Au meshes were used as current collectors for the anode and cathode, respectively.

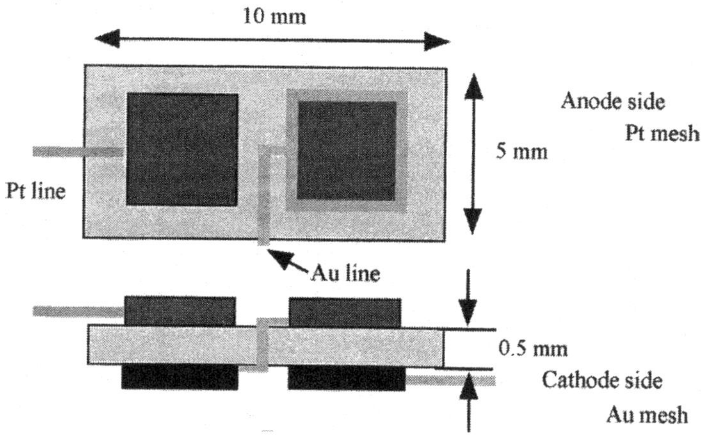

Figure 8: A schematic diagram of SC-SOFC module.

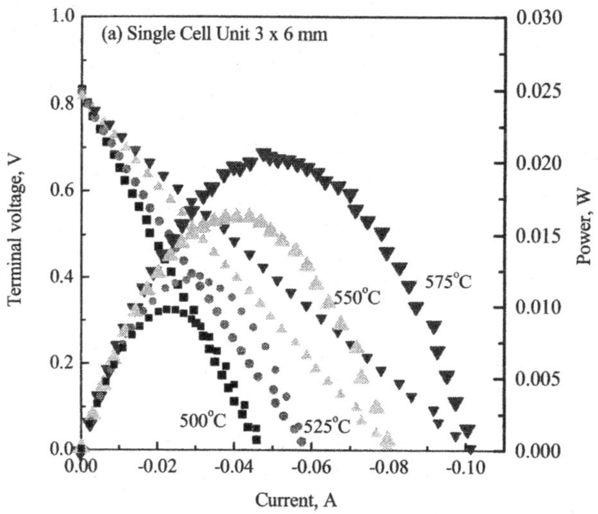

Figure 9a: I-V discharge profile for single cell module SC-SOFC (3x6 mm)

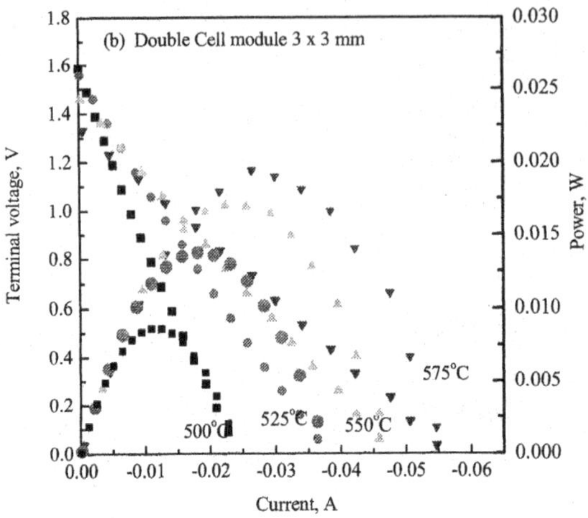

Figure 9b: I-V discharge profile for double cell module SC-SOFC (two 3x3 mm cels).

Figures 9a and 9b depict the discharge profiles of single and double cell configuration. The OCV of the single cell was stable and showed ~ 0.8 V within the test temperature range. A power density ~80 mW/cm^2 was obtained at 550°C. Considering the high electrolyte resistances (Fig. 6), the performance of the cell is limited by the electrolyte resistance. Hence, the cell performance can be improved by reducing the electrolyte thickness. On the other hand, the OCV obtained using the double cell module was as high as 1.6 V with a maximum current of 25 mA at 500°C.

Using the performance data above, a cell module for SC-SOFC is proposed. Figure 10 shows a schematic diagram of 6V-cell module inserted in a ceramic tube which serves as a gas flow path. This compact, 40 mm x 5 mmϕ, 6 V module would generate ~60 mW power at 550°C inlet gas temperature for various special applications. While SC-SOFC offers many advantages, such as sealant-free configuration and reasonably high power densities, the fuel utilization is a major drawback which will be reported in a future study.

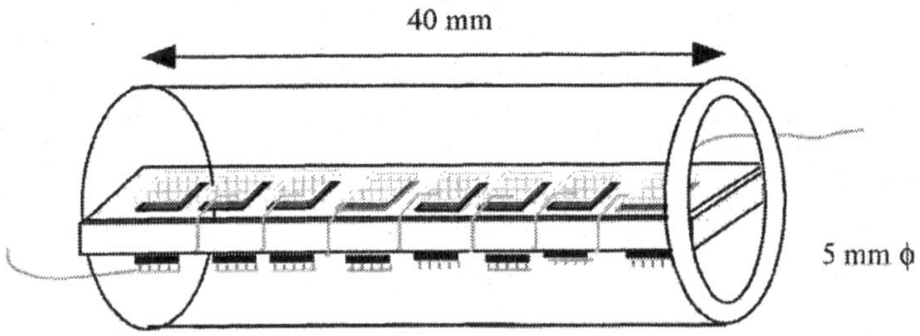

Figure 10: Schematic diagram of a proposed 6V-cell module.

SUMMARY

Influence of cathode material composition on the open circuit voltage (OCV) of single chamber solid oxide fuel cell was investigated. Cells with a mixture of SDC and (La, Sr)(Co, Fe)O_3 as cathode material and Ag reference electrode on SDC electrolyte were prepared and tested under single chamber conditions. Using LSCF as anode, the OCV was as high as 0.1 V, which decreased as the SDC content in LSCF increased. It was shown that LSCF exhibits a catalytic activity for oxidation of the fuel and lowers the OCV when it is used as an electrode in the single chamber SOFC. Measurements of SDC-LSCF composite cathode materials with Ni-cermet anode on SDC electrolyte showed that the OCV increased with increasing SDC content. Best performance among the tested samples was obtained with a power density 140 mWcm^{-2} at 600°C using 10SDC-LSCF as cathode material on a ~0.5 mm thick SDC electrolyte. These results were attributed to a strong influence of SDC content in LSCF on the cathode overpotential. Increasing amount of SDC in LSCF resulted in a decreasing overpotential of the cell.

The electrolyte supported two cell module (electrode area 3 x 3 mm) showed stable performance below 550°C furnace temperature. Based on these results, one would expect that a ceramic tube (4 cm length and 5 mmϕ) containing a 8 cell module with a single electrode area 3 x 3 mm would result in an open circuit voltage of 6 V and ~50 mA maximum current at 550°C. The power of such a cell can be further increased by reducing the thickness of the electrolyte.

ACKNOWLEDGEMENT

The authors would like to acknowledge a grant from the Research Board of University of Missouri-Rolla for the partial support of this work.

REFERENCES

[1] T. Hibino, A. Hashimoto, M. Yano, M. Suzuki, S. Yoshida, M. Sano "High performance anodes for SOFCs operating in methane-air mixture at reduced temperatures," *J. Electrochem. Soc.* 149 (2) A133-A136 (2002).

[2] T. Hibino, A. Hashimoto, T. Inoue, J. Tokuno, S. Yoshida, M. Sano "A solid oxide fuel cell using an exothermic reaction as the heat source," *J. Electrochem. Soc.* 148 (6) A544-A549 (2001).

[3] T. Hibino, A. Hashimoto, T. Inoue, J. Tokuno, S. Yoshida, M. Sano "A low-operating-temperature solid oxide fuel cell in hydrocarbon-air mixtures," *Science.* 288 p. 2031 (2000).

[4] R.J.H. Voorhoeve, in *Advanced Materials in Catalysis*, Edited by J.J. Burton and R.L. Garten, Academic Press, p.129 (1977).

[5] P. Jasinski, T. Suzuki, Z. Byars, F. Dogan, H.U. Anderson "Comparison of anode and electrolyte support configuration of single chamber solid oxide fuel cell," Electrochemical Society Proc. "Solid Oxide Fuel Cells VIII" PV 2003-07, (2003) 1101.

[6] P. Jasinski, T. Suzuki, X.D .Zhou, F. Dogan, H.U. Anderson "Single Chamber Solid Oxide Fuel Cell – Investigation of Cathode," Ceramic Engineering and Science Proceedings 24, 3, p.293(2003).

[7] P. Jasinski, T. Suzuki, F. Dogan and H.U. Anderson "Impedance Spectroscopy of Single Chamber SOFC," accepted *Solid State Ionics.*

PEROVSKITE TYPE BUFFERS FOR YBCO COATED CONDUCTORS

M. Parans Paranthaman, T. Aytug, H.Y. Zhai, A.A. Gapud, P.M. Martin, K.J. Leonard, A. Goyal, and D.K. Christen
Oak Ridge National Laboratory, Oak Ridge, TN 37831, USA

ABSTRACT

Coated conductor applications in power technologies require stabilization of the high-temperature superconducting (HTS) layers against thermal runaway. One route to stabilization is through the use of conducting buffer layers that provide good electrical and/or thermal connectivity throughout the laminate structure. Perovskite based conducting oxide buffers such as $LaNiO_3$ (LNO), $SrRuO_3$ (SRO), and $La_{0.7}Sr_{0.3}MnO_3$ (LSMO) have been developed for Rolling-Assisted Biaxially Textured Substrates (RABiTS) approach. Using sputtering, three multi-layer architectures such as SRO/LNO/Ni; LSMO/Ni/Cu; and LSMO/Ir/Ni-W were developed. High J_c $YBa_2Cu_3O_{7-\delta}$ (YBCO) films were grown on these conductive architectures using pulsed laser deposition. Detailed temperature dependent resistivity measurements indicate that YBCO/LSMO/Ir/Ni-W tape is fully conductive. A self-field J_c of 2.2 MA/cm^2 at 77 K was achieved for this tape. Both cross-sectional TEM and SIMS studies indicated the presence of clean substrate/buffer interface. These results demonstrate that LSMO/Ir multilayers may provide a fully conductive architecture for biaxially textured Cu and/or Cu-alloy substrates.

INTRODUCTION

The Rolling-Assisted Biaxially Textured Substrates (RABiTS) and Ion-Beam Assisted Deposition (IBAD) approaches have been identified as the leading techniques to fabricate 10-50 meter lengths of high performance $YBa_2Cu_3O_{7-\delta}$ (YBCO) coated conductors.[1-3] In the standard RABiTS approach, a three-layer buffer architecture of $CeO_2/YSZ/Y_2O_3$ is used to fabricate long lengths of buffered tapes by the epitaxial deposition on the thermo mechanically-textured Ni alloy substrates. Such a structure would need atleast 50 μm thick Cu stabilizers to provide stabilization to the superconductor during an over-current situation. On the other hand, the use of Cu substrates with fully conductive architectures would be desirable because they would enable reduced ferromagnetism, higher electrical conductivity, higher thermal conductivity, and lower cost compared to Ni and Ni-alloys. Hence, it is essential to develop a superconductor with a fully conducting laminate structure having a non-oxidizing protective buffer layer for the prevention of substrate oxidation and also diffusion of substrate components into the other buffers and/or superconductors, and also for providing stabilization to the superconductor during an over-current situation. This would also eliminate the need for Cu stabilizer layer on top of the superconductor, resulting in increase of the overall engineering current density.

Perovskite based oxide buffers play an important role in second generation YBCO coated conductors. The ideal ABO_3 perovskite structure is shown in Figure 1. In addition, the list of perovskite type oxides that have been developed for YBCO is also shown in Fig. 1. The material properties vary all the way from insulators, semiconductors to metals. Conducting buffers such as $LaNiO_3$ (LNO), $SrRuO_3$ (SRO), and $La_{0.7}Sr_{0.3}MnO_3$ (LSMO) have been explored as the potential candidates for this study. Using sputtering, several conductive multilayer architectures were developed. Here we report our successful demonstration of the fully conductive architectures for YBCO coated conductors.

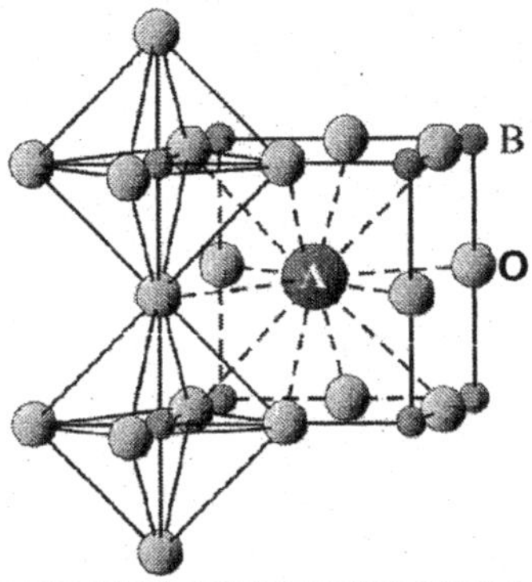

Ideal ABO_3 Perovskite Structure showing the Octahedral and 12-fold co-ordination of B and A-site cations

Material	Lattice parameter (pseudo cubic) Å	% lattice mismatch vs. YBCO	% lattice mismatch vs. Ni	Property
$BaCeO_3$	4.377	13.55	21.59	Insulator
$BaZrO_3$	4.193	9.27	17.34	Insulator
$SrRuO_3$	**3.941**	**3.08**	**11.17**	**Metal**
$SrTiO_3$	3.905	2.16	10.26	Insulator
$SrTi(Nb)O_3$	**3.905**	**2.16**	**10.26**	**Semiconductor**
$LaMnO_3$	3.880	1.60	9.70	Insulator
$(LaSr)MnO_3$	**3.860**	**0.98**	**9.07**	**Metal**
$LaNiO_3$	**3.859**	**0.98**	**9.07**	**Metal**
$NdGaO_3$	3.841	0.51	8.61	Insulator
$LaAlO_3$	3.793	-0.75	7.35	Insulator

FIGURE 1. THE IDEAL ABO_3 PEROVSKITE STRUCTURE. TABLE PROVIDES THE LIST OF OXIDE BUFFERS DEVELOPED FOR YBCO COATED CONDUCTORS.

EXPERIMENTAL PROCEDURE

The SRO/LNO multi-layers were grown on biaxially textured Ni substrates using rf sputtering. The sputter targets used for these experiments were made by lightly packing the single phase SRO and LNO powders into a 4" copper tray. LNO layers were deposited at temperatures ranging from 450 to 500 °C in the presence of Ar/H_2 4% gas mixtures. The deposition rate was approximately 0.042 nm/sec, and the resulting thickness was 300 nm. The details of the experimental conditions were published earlier.[4,5] Subsequent growth of SRO layers on LNO-buffered Ni substrates was performed at a temperature ranging from 600 to 650 °C under 10 mTorr Argon. The deposition rate for SRO was 0.048 nm/sec with the resulting thickness of 400 nm.

The LSMO/Ni multi-layers were grown on biaxially textured Cu substrates. To minimize the Cu diffusion and also to reduce the oxidation rate of the substrate, a protective 1.6 μm thick Ni overlayer was deposited by dc sputtering at a temperature of 600 °C in a reducing atmosphere of forming gas (Ar/H_2 4%). Subsequent growth of the LSMO layers on Ni-buffered Cu substrates was performed at a temperature of 550 to 625 °C using rf sputtering in a mixture of forming gas and 2×10^{-5} Torr of H_2O. The typical LSMO thickness is around 300 nm. The details of the experimental conditions were published earlier.[6]

The LSMO/Ir multilayers were grown on biaxially textured Ni-W3 at.% substrates. Ir metal is identified as having excellent oxidation and corrosion resistance among platinum group elements. In addition, its native oxide, IrO_2 is also a conducting layer with low oxygen diffusivity. These properties make Ir a unique candidate for this study. Ir thin films were deposited on Ni-W substrates using dc sputtering at a temperature of 550 to 650 °C in the presence of $Ar\text{-}H_2$ 4%. Subsequent LSMO deposition was performed at 625 to 700 °C in the presence of forming gas and 5×10^{-5} Torr H_2O. The details were published earlier.[7]

YBCO films were deposited on these three multi-layers by pulsed laser deposition at 780 °C in 120 m torr oxygen with an average laser energy of 400-410 mJ using a stoichiometric YBCO target, followed by annealing under 550 Torr oxygen during cool down. Typical YBCO thickness was varied from 200 to 700 nm. The crystalline structure of the films was analyzed by X-ray diffraction techniques. SEM micrographs were taken using a Hitachi S-4100 field emission microscope. Cross-section examination of the samples were conducted by a high resolution transmission electron microscope (TEM) equipped with energy dispersive spectrometry (EDS) unit. The films were then prepared for current density measurements by depositing silver for current and voltage leads, followed by oxygen annealing at 500 °C for 1 h. The transport critical current density, J_c, was measured using a standard four-point probe technique with a voltage criterion of 1 μV/cm.

RESULTS AND DISCUSSION

$LaNiO_3$ (LNO) buffers were grown epitaxially on biaxially textured Ni substrates using sputtering. YBCO films grown on LNO-buffered Ni had a reduced T_c and also poor superconducting properties. This is mainly due to the diffusion of Ni from the substrate into the superconductor. Also, $SrRuO_3$ (SRO) buffers were grown untextured on Ni substrates. However, highly textured SRO layers were grown on LNO-buffered Ni substrates. YBCO films with a J_c of 1.3 MA/cm^2 for a thickness of 200nm have been achieved on conducting architecture of SRO/LNO/Ni substrates. The field dependent J_c for this tape is shown in Fig. 2. The irreversibility field for this tape is 7.5 T. The temperature dependent net resistivity for YBCO/SRO/LNO/Ni conducting structure is shown in Figure 3. YBCO film showed a metallic behavior from Room Temperature (RT) down the transition temperature (T_c) of 91 K. The measured room temperature net resistivity is around 15 $\mu\Omega$-cm. This indicates that there is some contact barrier between the superconductor and Ni substrate through SRO/LNO. SIMS, SEM, and X-ray indicated the presence of non-conducting impurities at the buffer/Ni interface. However, NiO impurities were not observed before the YBCO deposition. As can be seen from Fig. 3, SRO/LNO layers are fully connected to the underlying Ni substrate.

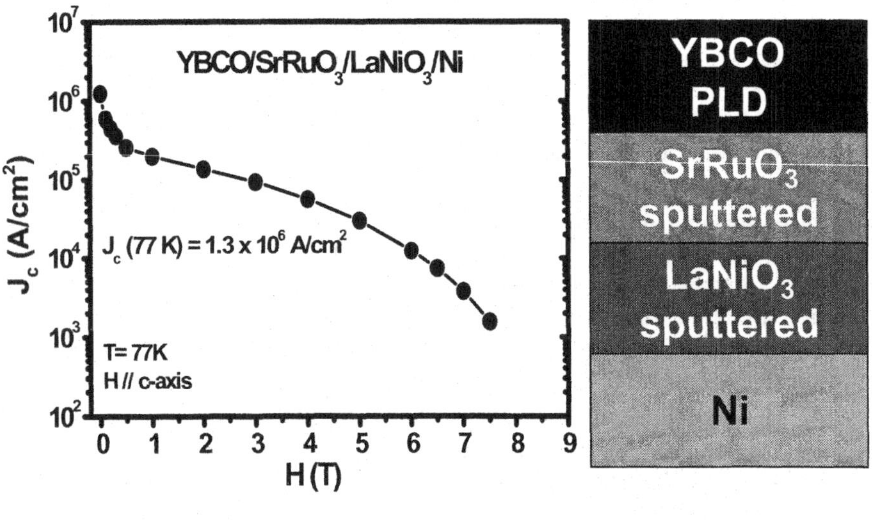

FIGURE 2. THE FIELD-DEPENDENCE OF THE TRANSPORT CRITICAL CURRENT DENSITY, J_c FOR A 200 nm THICK PLD YBCO FILM ON SPUTTERED $SrRuO_3/LaNiO_3$ MULTI-LAYERS ON BIAXIALLY TEXTURED Ni SUBSTRATES.

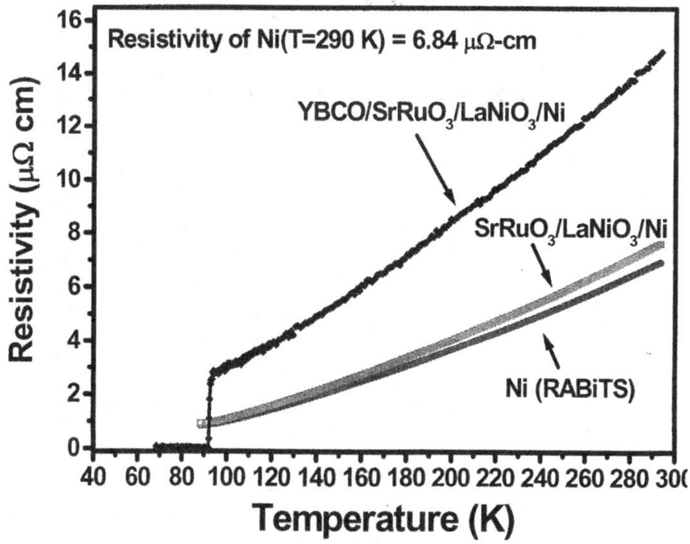

FIGURE 3. TEMPERATURE-DEPENDENCE OF THE NET RESISTIVITY FOR YBCO/$SrRuO_3$/$LaNiO_3$/Ni TAPE SHOWN IN FIGURE 2. ALSO SHOWN FOR COMPARISON ARE THE ρ_{net}-T DATA FOR AN AS-GROWN $SrRuO_3$/$LaNiO_3$/Ni TAPE AND FOR A BIAXIALLY TEXTURED Ni SUBSTRATE.

Ni over-layers were grown epitaxially on biaxially textured Cu substrates using sputtering. Highly textured $La_{0.7}Sr_{0.3}MnO_3$ (LSMO) layers were then grown on Ni-buffered Cu substrates. YBCO films with a high J_c of 2.3 MA/cm^2 were grown on conductive LSMO/Ni/Cu structures. The field dependent J_c for this conductive structure is shown in Fig. 4. The measured irreversibility for this tape is around 7.5 T. The temperature dependent resistivity for YBCO/LSMO/Ni/Cu tape is shown in Figure 5. The room temperature resistivity for this tape is around 6 μΩ-cm and the measured T_c is 90.4 K. SIMS analyses indicate that LSMO layer is a good Ni diffusion layer. However the NiO impurity was also observed at the Ni/LSMO interface. Also, Cu and Ni layers are completely miscible and Cu/Ni interdiffusion depends mainly on the YBCO process temperatures. That is the reason thick Ni overlayers were deposited. However, the presence of thick Ni overlayer contributes to the ferromagnetism in the tape. Hence, there is a need to develop a fully conductive architecture based on non-magnetic overlayers. For comparison, resistivity data for both LSMO/Cu and Cu tapes are also plotted in Fig. 5.

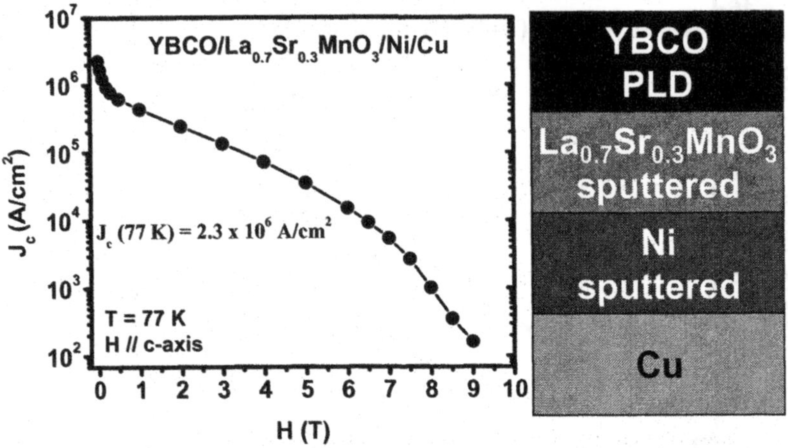

FIGURE 4. THE FIELD-DEPENDENCE OF THE TRANSPORT CRITICAL CURRENT DENSITY, J_c FOR A 200 nm THICK PLD YBCO FILM ON SPUTTERED $La_{0.7}Sr_{0.3}MnO_3$/Ni MULTI-LAYERS ON BIAXIALLY TEXTURED Cu SUBSTRATES

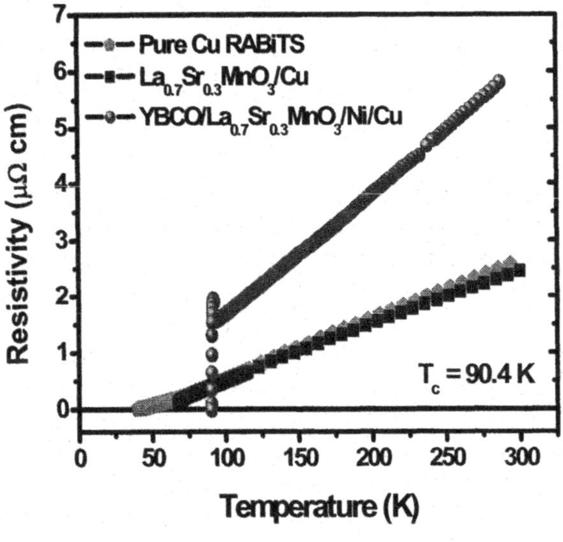

FIGURE 5. TEMPERATURE-DEPENDENCE OF THE NET RESISTIVITY FOR YBCO/$La_{0.7}Sr_{0.3}MnO_3$/Ni/Cu TAPE SHOWN IN FIGURE 4. ALSO SHOWN FOR COMPARISONARE THE ρ_{net}-T DATA FOR AN AS-GROWN $La_{0.7}Sr_{0.3}MnO_3$/Cu TAPE AND FOR A BIAXIALLY TEXTURED Cu SUBSTRATE.

Ir buffers were grown epitaxially on biaxially textured Ni-W substrates. LSMO buffers were then grown on Ir-buffered Ni-W substrates. YBCO films with a J_c of 2.2 MA/cm^2 (for 200 nm thick) and a J_c of 1.3 MA/cm^2 (for 700 nm thick) were grown on conductive LSMO/Ir/Ni-W tapes. The field dependent J_c for this conductive structure is shown in Fig. 6. The irreversibility fields for 200 nm thick YBCO film are 6.8 T and for 700 nm thick YBCO films are 5.8 T. The temperature dependent resistivity for these conductive architectures is shown in Fig. 7. For comparison, resistivity data for Ni-W and LSMO/Ir/Ni-W tapes are also shown in Figure 7. These resistivity data indicate that YBCO films are completely connected to the underlying Ni-W substrate. This is the first demonstration of the fully conductive architectures developed for second generation YBCO coated conductors. The observed T_c for these conductive structures is 88-89 K. The cross-sectional TEM bright field image for 200 nm thick YBCO film on fully conductive LSMO/Ir/Ni-W substrates is shown in Figure 8. All the interfaces are clean and there is no formation of oxide impurities at the substrate/buffer interface. The selected area diffraction data taken at all the interfaces indicate that YBCO, LSMO, Ir films are cube-on-cube oriented. It was found that Ni-W and Ir also interdiffuses under certain conditions. Details are currently under investigation. Efforts are being made to grow conductive LSMO/Ir buffers on textured Cu substrates and the results will be published later.

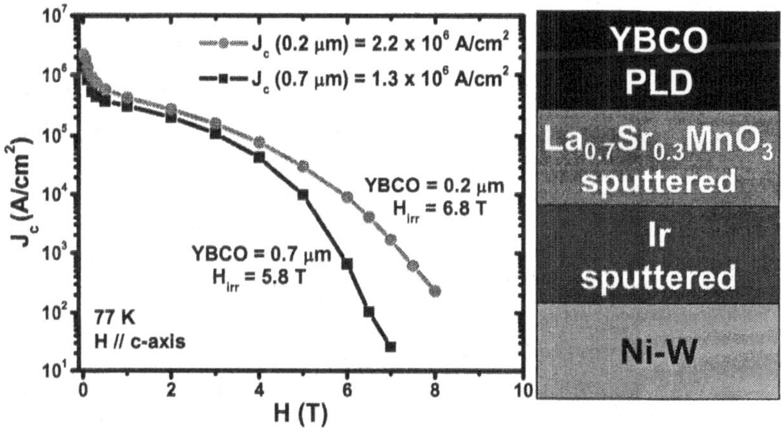

FIGURE 6. THE FIELD-DEPENDENCE OF THE TRANSPORT CRITICAL CURRENT DENSITY, J_C FOR A 200 nm THICK PLD YBCO FILM ON SPUTTERED $La_{0.7}Sr_{0.3}MnO_3$/Ir MULTI-LAYERS ON BIAXIALLY TEXTURED Ni-W SUBSTRATES

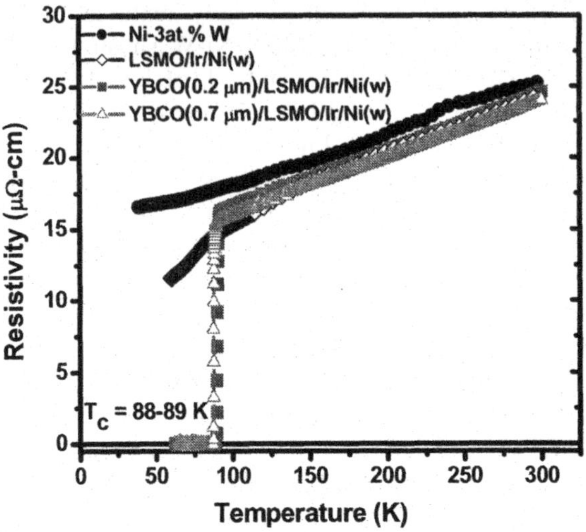

FIGURE 7. TEMPERATURE-DEPENDENCE OF THE NET RESISTIVITY FOR TWO YBCO/$La_{0.7}Sr_{0.3}MnO_3$/Ir/Ni-W TAPES SHOWN IN FIGURE 6. ALSO SHOWN FOR COMPARISON ARE THE ρ_{net}-T DATA FOR AN AS-GROWN $La_{0.7}Sr_{0.3}MnO_3$/Ir/Ni-W TAPE AND FOR A BIAXIALLY TEXTURED Ni-W SUBSTRATE.

SUMMARY

We have developed three conductive architectures of YBCO/SRO/LNO/Ni; YBCO/LSMO/Ni/Cu; and YBCO/LSMO/Ir/Ni-W for second generation YBCO coated conductors. Detailed temperature dependent resistivity data indicate that fully conductive architectures were observed for YBCO/LSMO/Ir/Ni-W tapes. A self-field J_c of 2.2 MA/cm^2 was observed for this conductive structure. These results demonstrate that fully conductive architectures of YBCO/LSMO/Ir may be suitable for biaxially textured Cu or Cu-alloy substrates.

ACKNOWLEDGEMENTS

This work was supported by the U.S. Department of Energy, Division of Materials Sciences, Office of Science, Office of Electric Transmission and Distribution. This research was performed at the Oak Ridge National Laboratory, managed by U.T.-Battelle, LLC for the USDOE under contract DE-AC05-00OR22725.

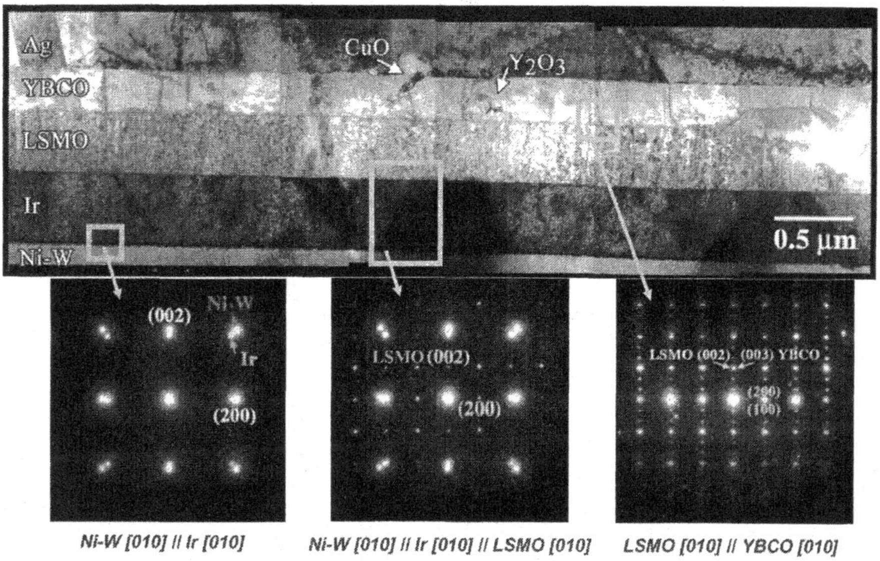

FIGURE 8. CROSS-SECTIONAL TEM BRIGHT-FIELD IMAGE OF THE 200 nm THICK YBCO FILM ON $La_{0.7}Sr_{0.3}MnO_3$/Ir BUFFERED Ni-W TAPE. ALL THE Ni-W/Ir, Ir/LSMO, LSMO/YBCO INTERFACES ARE CLEAN. THERE IS NO OXIDE PRESENT AT THE Ni-W/Ir INTERFACE. THE SELECTED AREA DIFFRACTION PATTERN TAKEN AT ALLTHE INTERFACES SHOW ALL THE LAYERS CUBE-ON-CUBE ORIENTED.

REFERENCES

[1]A. Usoskin, H.C. Freyhardt, A. Issaev, J. Dzick, J. Knoke, M.P. Oomen, M. Leghissa, and H-W. Neumueller, "Large area YBCO-coated stainless steel tapes with high critical currents," IEEE Trans on Appl. Supercond. **13**, 2452 (2003).

[2]Y. Iijima, K. Kakimoto, and T. Saitoh, "Fabrication and transport characteristics of long length Y-123 coated conductors processed by IBAD and PLD," IEEE Trans. on Appl. Supercond. **13**, 2466 (2003).

[3]D.T. Verebelyi, U. Schoop, C. Thieme, X. Li, W. Zhang, T. Kodenkandath, A.P. Malozemoff, N. Nguyen, E. Siegal, D. Buczek, J. Lynch, J. Scudiere, M. Rupich, A. Goyal, E.D. Specht, P. Martin, and M. Paranthaman, "Uniform performance of continuously processed MOD-YBCO-coated conductors using a textured Ni-W substrate," Supercond. Sci. and Technol. **16**, L19 (2003).

[4]T. Aytug, J.Z. Wu, C. Cantoni, D.T. Verebelyi, E.D. Specht, M. Paranthaman, D.P. Norton, D.K. Christen, R.E. Ericson, and C.L. Thomas, Appl. Phys. Lett. **76**, 760 (2000).

[5]T. Aytug, B.W. Kang, C. Cantoni, E.D. Specht, M. Paranthaman, A. Goyal, D.K. Christen, D.T. Verebelyi, J.Z. Wu, R.E. Ericson, C.L. Thomas, C-Y. Yang, and S.E. Babcock, J. Mater. Res. **16**, 2661 (2001).

[6]T. Aytug, M. Paranthaman, J.R. Thompson, A. Goyal, N. Rutter, H.Y. Zhai, A.A. Gapud, A.O. Ijaduola, and D.K. Christen, Appl. Phys. Lett. **83**, 3963 (2003).

[7]T.Aytug, M. Paranthaman, H.Y. Zhai, A.A. Gapud, K.J. Leonard, J.R. Thompson, and D.K. Christen, Appl. Phys. Lett. (in press).

POLARIZATION DYNAMICS OVER BROAD TIME AND FIELD DOMAINS IN MODIFIED FERROELECTRICS

Christelle Jullian, Jiefang Li and D. Viehland,
Dept. of Materials Science and Engineering, Virginia Tech
Blacksburg, VA 24061

ABSTRACT

The dynamics of polarization switching have been investigated over extremely broad time ($10^{-8}<t<10^{2}$ sec) and field ranges for various modified Pb-based perovskite ferroelectrics. The results unambiguously demonstrate the presence of extremely broad relaxation time distributions for switching, which can extend over decade(s) in orders of magnitude in time.

INTRODUCTION

Limiting the study of domain dynamics and polarization switching has been that current transient investigations have been performed over relatively narrow time (t) and electric field (E) ranges [1-8] – even though the current response is known to be logarithmic in time [4, 9-11]. Analysis of the dynamics in the time domain of $10^{-8}<t<10^{-6}$ sec provides incomplete information, upon which to develop a mechanistic understanding. For example, it is known that the current transient peaks are skewed to longer times in the domain of $10^{-8}<t<10^{-6}$ sec [4,6-8]. Skewing has been understood by (i) a geometric model that includes domain impingement [4]; and (ii) stretched exponential functions that include the effect of random fields which broaden nucleation and growth events in the time domain [10]. Also, recent P-E studies of modified Pb-based perovskites have revealed pronounced dispersion in the frequency domain of $10^{-2}<f<10^{2}$ Hz [11]. This dispersion has also been shown to follow stretched exponential functions.

Stretched exponential behavior is well-known in disordered systems [12-14]. It is a form of hierarchial relaxation. Many modified perovskite ferroelectrics are also known to contain significant disorder, such as $(1-x)Pb(Mg_{1/3}Nb_{2/3})O_3-(x)PbTiO_3$ (PMN-x%PT) [15-17]. These systems posses random-fields [14,18], which couple to the polarization [19]. Random-fields have also been shown to be important to the piezoelectric behavior of "soft" $Pb(Zr_{1-x}Ti_x)O_3$ (PZT), which are modified with higher valent substituents [20,21,11], such as La^{+3} on the Pb^{+2} site (i.e., PLZT).

Investigations of the polarization dynamics over broad time and field regions would be greatly important to the study of modified ferroelectrics. It could significantly impact understanding of domain dynamics. In this letter, we report the polarization dynamics over such a broad time domain, extending from $10^{-8}< t <10^{2}$ sec. We have studied various modified Pb-based ferroelectrics. Studies were performed from $E<<E_c$ to $E>>E_c$, where E_c is the coercive field. The results unambiguously demonstrate the presence of extremely broad relaxation time distributions for the switching process, extending over decade(s) in orders of magnitude in time, where the distribution is strongly dependent on E.

EXPERIMENTAL PROCEDURE

Polycrystalline specimens of PMN-PT 70/30 (E_c~4 kV/cm) and PLZT 7/65/35 (E_c~6.7 kV/cm) were prepared, as previously reported [21]. Soft PZT (E_c~11 kV/cm) specimens were obtained from EDO Corp. (Salt Lake City, UT). In addition, a (001)-oriented single crystal of PMN-PT 70/30 was obtained from HC Materials (Urbana, IL). The specimens were cut into typical dimensions of 0.3 mm in thickness and 4 mm^2 in area, and were electroded with gold. Measurements were performed for various fields of $0<E<3E_c$. In order to measure the response of the specimens over a broad time domain from $10^{-8}<t<10^{2}$ sec, three different measurements circuits were developed and built [22]. Figure 1(a) shows the measurement circuit for the short-time domain between 10^{-8} to 10^{-6} sec. Figure 1(b) shows the circuit for the middle-time domain between 10^{-6} to 10^{-3} sec. And, Figure 1(c) shows the circuit for the long-time domain between 10^{-3} to 10^{2} sec. For the short and middle time domains, the same procedure was used prior to switching. Initially the specimen was unpoled, and then an electric field of $-3E_c$ was applied to re-pole the specimen. To prepare for switching, both sides of the specimen were raise to the desired switching field. Then, at time t=0, one side of the specimen was taken to ground: this is an important step that allowed for the very highest current wall between the power amplifier and the specimen. By using these steps, we can certify that high voltage reaches its full maximum, before the polarization starts to rise. It was found that switching times of $<<10^{-8}$ seconds could be achieved. For the long-time domain, this method could also be used, but is not necessary as the rise time of the amplifier is significantly shorter than the measurement time. An Agilent oscilloscope operated in a time capture mode was used to measure the output voltage from each circuit.

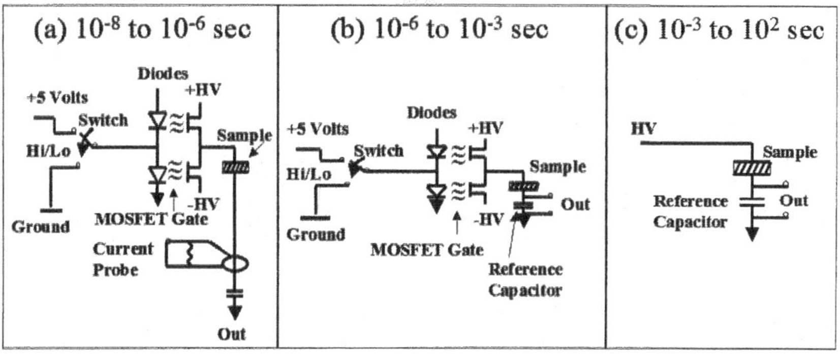

Figure 1. Measurement circuit of switching current responses for the various time domains: (a) short time domain between $10^{-8}<t<10^{-6}$ seconds; (b) middle time domain between $10^{-6}<t<10^{-3}$ seconds; and (c) long time domain between $10^{-3}<t<10^{2}$ seconds. The circuits behind the specimen are equivalent for the short and middle time domains. This circuit consists of a switch, rapid response diodes, and MOSFET gates. In the fast time domain, a current probe is used to lower the input voltage into an oscilloscope. For the middle and long time domains, the circuits in front of the specimen are equivalent. The voltage is measured across a reference capacitor, using an oscilloscope operated in a time capture mode.

RESULTS

Figure 2 shows the logarithm of the polarization as a function of the logarithm of time for (a) a PMN-PT 70/30 ceramic, (b) PLZT 7/65/35, (c) PZT, and (d) a (001)-oriented PMN-PT 70/30 crystal. All three of these specimens are soft ferroelectrics at room temperature, which have a modest E_c and relatively high susceptibilities. Data are shown for over ten decades in time, taken at various applied electric fields of $3<E<20$ kV/cm. The data can be seen to be quite similar for all three compositions. It is similar for single crystals and polycrystals. Thus, the interesting general features of the data for the various compositions will be discussed together below.

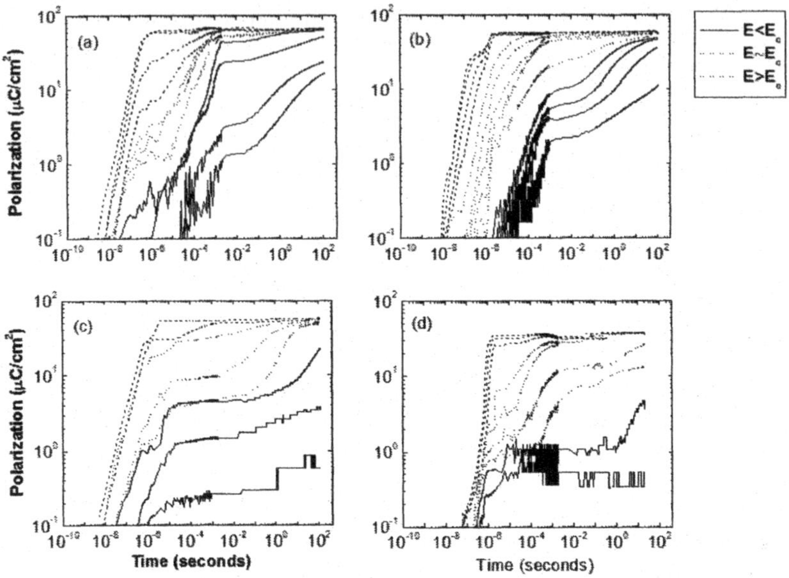

Figure 2. Logarithm of polarization as a function of logarithm of time for various "soft" ferroelectrics over broad time and field ranges. (a) PLZT 7/65/35, (b) PMN-30%PT, and (c) a commercial "soft" PZT. Data are shown for 3 different field regimes of $E<E_c$, $E\approx E_c$, and $E>E_c$ regimes in each figure.

The spectra can clearly be seen to be extremely broad in the time domain, extending over decade(s) of orders in magnitude. The spectra only became sharp at higher fields of $E>>E_c$ and short times of $t<10^{-6}$ seconds. Unambiguously, polarization switching in soft ferroelectrics has a very broad distribution of relaxation times τ. It is important to note that the results for the PMN-PT ceramic where quite similar to those of the (001)-oriented crystal, demonstrating that the broadness in the time-domain is not due to a variety of extrinsic pinning sites such as grain boundaries and dislocations. The breadth of this distribution is nearly equal to that found in the weak-field susceptibility (dielectric constant) of relaxor ferroelectrics near their freezing temperature [17,23,24]. However, there is structure within the spectra that changes with E and log(t). Below we use PLZT 7/65/35 as an example by which to discuss these features.

For $E<E_c$, as shown in Figure 3a, two regimes of different polarization responses were found. At short times of $10^{-5}<t<10^{-3}$ sec, the polarization logarithmically increased with t, exhibiting significant fluctuation (Barkhaussen-like [14]) events. With increasing E the randomness of the fluctuations decreased, and a plateau was found for $10^{-3}<t<10^{-1}$ sec. At longer times of $10^{-1}<t<10^{2}$ sec, the polarization smoothly and gradually increased with log(t). For $E<E_c$, polarization saturation seemingly would eventually be achieved, but only at times many orders of magnitude longer than 10^2 sec. For $E\approx E_c$, two regimes of polarization response were again present, as shown in Figure 3b. With increasing E, the regimes became increasingly indistinguishable in the time domain of $10^{-7}<t<10^{-3}$ sec. Complete saturation was only gradually approached for $t>10^{-3}$ sec. For $E>E_c$, the polarization response became increasingly sharp with increasing E, as shown in Figure 3c. For $E>>E_c$, saturation was reached at $\sim 10^{-7}$ sec.

We discuss our results in terms of a double well potential that is under an applied electric field E and that has a rugged fine-structure, which is somewhat like that of a relaxor [17,23,24]. For small E, Barkhaussen-like events occur as the system moves between rugged fine structures, towards the other variant – a polarization creep develops over long times. This creep occurs by stretched exponential functions [10,11], due to the hierarchial nature of relaxation within a rugged landscape [12,13]. With increasing E, the system moves more freely within this landscape, and thus the breadth of the τ distribution and $\tau_{average}$ both decrease. At high fields of $E>>E_c$, polarization switching occurs rapidly and the τ distribution is sharp. In this field range, the dynamics can be described by an Avarami-type relationship [1-8], as the rugged landscape presents no impediments to switching.

Previous electron microscopy studies of soft ferroelectrics [20,21] have shown domain breakdown with increasing AC electric field. Furthermore, fractal domains have recently been used to understand long-time polarization dynamics in various modified PZTs [10,11,25]. The results of our investigation demonstrate that soft ferroelectrics contain such heterogeneities under field for $E\leq E_c$, which control the dynamics of polarization switching. These heterogeneities are conceptually similar to the long-lived ones observed in relaxor ferroelectrics under weak field drive. At higher fields of $E>>E_c$, switching occurs by a mechanism similar to conventional ferroelectric domain nucleation and growth [1-8].

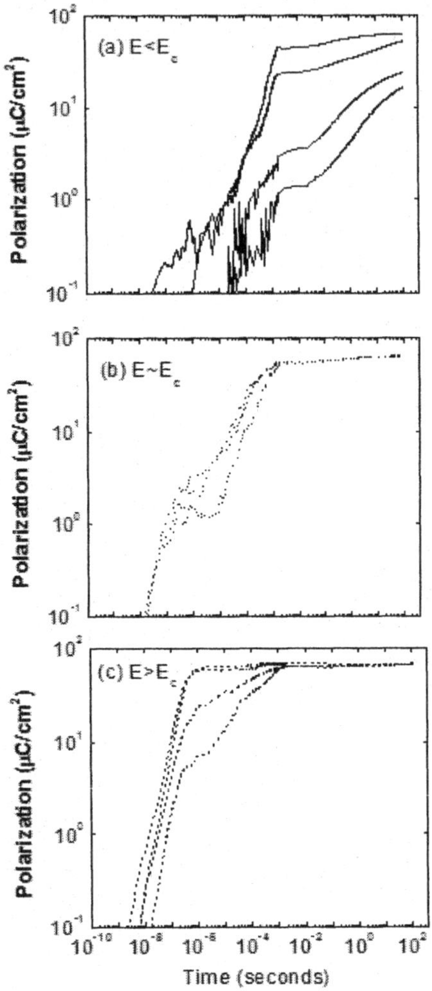

Figure 3. Logarithm of polarization as a function of logarithm of time for PLZT 7/65/35 for various field ranges. (a) $E<E_c$ [3, 4, 5, and 6 kV/cm], (b) $E \approx E_c$ [6.4, 6.7 7 kV/cm], and (c) $E>>E_c$ [10, 15, 17, and 19 kV/cm]. The fields shown in the brackets represent data from lower right to upper left in each figure.

SUMMARY

In summary, the polarization response of various modified Pb-based perovskites has been investigated over broad time and field ranges. The results unambiguously demonstrate the presence of extremely broad relaxation time distributions for switching, which can extend over decade(s) in orders of magnitude in time. With increasing field for $E>>E_c$, the breadth of the distribution is significantly sharpened, revealing that the switching mechanism is dependent on E.

ACKNOWLEDGEMENTS

This research was supported by the Office of Naval Research under grants N000140210340, N000140210126, and MURI N000140110761.

REFERENCES

[1]W.J. Merz, *Physical Review* 95, 690-698 (1954).
[2]M.E. Drougard, *J. Appl. Phys.* 31, 352-355 (1960).
[3]R.C. Miller and G. Weinreich, *Phys. Rev.* 117, 1460-1466 (1960).
[4]V. Shur, E. Rumyantsev, and S. Makarov, *J. Appl. Phys.* 84, 445-451 (1998).
[5]Y. Ishibashi and Y. Takagi, *J. Phys. Soc. Japan* 31, 506-511 (1971).
[6]A. Levstik, M. Kosec, V. Bobnar, C. Filipic, and J. Holc, *Jap. J. Appl. Phys.* 36, Pt. 1, No. 5A, 2744-2746 (1997).
[7]Y. Ishibashi, *Integrated Ferroelectrics* 2, 41 (1992).
[8]T. Song, S. Aggarwal, Y. Gallais, B. Nagaraj, R. Ramesh, and J. Evans, *Appl. Phys. Lett.* 73, 3366-3368 (1998).
[9]I. Boscolo and S. Cialdi, *J. Appl. Phys.* 91, 6125 (2002).
[10]D. Viehland and J.F. Li, *J. Appl. Phys.* 90, 2995 (2001).
[11]D. Viehland and Y. Chen, *J. Appl. Phys.* 88, 6696 (2000).
[12]J. Mydosh, in "*Spin glasses, An experimental introduction*", Taylor and Francis, London (1993).
[13]S. Hutton, U. Hochli, M. Magilone, in "*Relaxation in Complex Systems and Related Topics*", ed. I. Campbell and C. Biovannella, (Plenum, NY), 289 (1990).
[14]W. Kleeman, *Intern. J. Mod. Physics* 7, 2469 (1993).
[15]G.A. Smolenskii and A. Agranovskaya, *Sov. Phys. Sol. State* 1, 1429-1441 (1960).
[16]L.E. Cross, Ferroelectrics 151, 305-315 (1994).
[17]D. Viehland, M. Wuttig, and L.E. Cross, *J. App. Phys.* 68, 2916-2924 (1991).
[18]D. Viehland, L.E. Cross, J. Powers, and J.F. Li, *Appl. Phys. Lett.* 78, 3508-3510 (2001).

[19]A.P. Levanyuk and A.S. Sigov, in "*Defects and Structural Phase Transitions*", Gordon and Breach Science Publishers, New York, New York (1988).
[20]Qi Tan and Dwight Viehland, *Phys. Rev. B* 53, 14103-14111 (1996).
[21]Qi Tan, *PhD Dissertation*, University of Illinois, Urbana, IL (1998).
[22]We acknowledge that the control units and high-voltage test heads were designed and built by Paul Moses, State College, PA 16801.
[23]D. Viehland, M. Wuttig, and L.E. Cross, *Phil. Mag. B* 64, 335-344 (1991).
[24]D. Viehland, M. Wuttig, and L.E. Cross, *Ferroelectrics* 120, 71-77 (1991).
[25]S. Priya, H.Y. Kim, J. Ryu, K. Uchino, and D. Viehland, *Appl. Phys. Lett.*, 80(3) (2002).

$Ba(Zr_xTi_{1-x})O_3$:MgO COMPOSITES FOR FIELD AND FREQUENCY TUNABLE APPLICATIONS

George Y. Li*#, Edward Alberta, Zhi Yu, Ruyan Guo,# and A.S. Bhalla#

Materials Research Institute and #Department of Electrical Engineering
The Pennsylvania State University, University Park, PA 16802

ABSTRACT

To modify the dielectric constant and to improve loss characteristics of the ferroelectric $Ba(Zr_xTi_{1-x})O_3$ (BZT), BZT:MgO ceramic composites are made with dielectric MgO as the discrete phase in BZT matrix. This paper reports dielectric and microstructural studies of the BZT:MgO composites, with 25% and 30% Zr, and at 50 wt% MgO. Dielectric permittivity and loss tangent, electric field dielectric tunability and the K-factor (figure of merit) as function of frequency, temperature, and bias electric field conditions are reported and compared with phase pure BZT. MgO discrete phase is shown to be beneficial to decrease dielectric permittivity and loss, thus to optimize the figure of merit for tunable microwave electronic applications.

INTRODUCTION

Electric field tunable materials have been studied extensively in recent years due to their potential applications in tunable filters, antennas, phase shifters, etc. These applications require a material that demonstrates high tunability, low dielectric loss, and desired permittivity.[e.g., 1] Electric tuning is defined by using an external electrical field to alter the RF characteristics of the material, such as its capacitance. Multiple filterbanks, for an example, may be replaced with a single tunable filter, using electric field agile dielectric material such as the modified BZT:MgO. As a result, there has been great interest in materials that are tunable with an electric field.

$BaTiO_3$ is a well-known ferroelectric material that has been widely studied, but when used as a microwave material it suffers from high dielectric constant, temperature dependence of dielectric constant, high dielectric loss, a destructive phase transition near room temperature, and is difficult to have reproducible performance (due to large thermal hysteresis). Partially substituting Zirconium in place of Titanium yields BZT, a solid-solution system, in which the phase transition temperatures vary with the Zr/Ti ratio.[2,3] Using suitable amount of Zr to replace Ti in the solid solution yields pinched phase transition, which lowers the permittivity and helps the material's dielectric properties become less temperature dependent, by obtaining the paraelectric phase at room temperature and by bring other ferroelectric phase transitions into a narrower temperature range. Manipulating the Zr concentration in BZT yields tailored or controlled properties, which are useful in an array of different applications.[4,5]

To further enhance the properties of BZT a composite system is designed by introducing MgO, a well-known microwave material with low absorption, into BZT, similar as in the case of $(Ba,Sr)TiO_3$:MgO system.[67] This is done to modify the high dielectric constant and loss

* Participant of the NSF Research Experience for Undergraduates in Electrical Engineering at Penn State University, Summer 2003.

characteristics of BZT to levels suitable for microwave devices. With the introduction of MgO, along with its low dielectric constant and loss, a suitable permittivity (100-400) may be obtained. The estimation can be done using the following dielectric mixing rule:

$$\log\kappa_{eff} = \log v_1\kappa_1 + \log v_2\kappa_2 + \cdots + \log v_N\kappa_N \qquad (1)$$

where κ_{eff} is the effective permittivity of the composite, κ_N is the permittivity of the component N, and v_N is the volume fraction of component N.

It is of special interests to investigate the composite's cryogenic properties that will help determine its feasibility in space and its integration with superconductor device applications.

BZT is also of high interest because it is a lead-free compound. Environmental concerns are a contributing factor driving the exploration for lead-free piezoelectric and dielectric materials.

EXPERIMENTAL PROCEDURE

The BZT-MgO composites were prepared using conventional ceramic processing methods similar as that described in a previous report.[3] The compositions studied were $Ba(Zr_{0.25}Ti_{0.75})O_3$:50wt%MgO [BZT25:MgO] and $Ba(Zr_{0.30}Ti_{0.70})O_3$:50wt%MgO [BZT30:MgO].

The well-sintered ceramic is polished to obtain parallel plate sample, with sputtered gold electrode and attached silver wires.

Dielectric constant and loss measurements were made using an automated measurement system. This system consists of an LCR meter (HP-4284, Hewlett-Packard Inc.), closed–cycle helium cooled cryostat (HC-2, APD Cryogenics Inc.), temperature controller (model 330, Lakeshore Inc.), vacuum pump (Drytel 31, Alcatel), and a desktop computer. Dielectric measurements were made at cryogenic temperatures up to room temperature in the range of 13K to 300K at a rate of 2K/min, for discrete frequencies of 0.1 to 1MHz. Tunability was measured by measuring the dielectric constant of the sample under an applied electric field using a voltage source (TREK 610, Trek Inc.) and a high voltage blocking circuit, which effectively separates the LCR meter from the voltage (0 to 2kV) applied to the sample. The tunability is determined by finding the change in dielectric constant at zero-field compared to those dielectric values when there is an applied electric field at a given temperature, using the equation:

$$\text{Tunability} = \frac{\kappa(0) - \kappa(E)}{\kappa(0)} \qquad (2)$$

where $\kappa(0)$ is the zero-field dielectric constant and $\kappa(E)$ is the dielectric constant that results from an applied electric field of E. The tunability is most commonly expressed as a percentage.

The dielectric loss data is also collected at each condition of temperature and field. By using equation (2), and the values of loss tangent, the K-Factor

$$\text{K-factor} = \frac{Tunability}{\tan\delta} \qquad (3)$$

was determined and plotted as a function of temperature.

RESULTS AND DISCUSSION

1.Dielectric Behavior without Bias Electric Field

The dielectric properties were measured for the BZT:MgO composites containing Zr concentration of 25% and 30% in the broad range of 100Hz to 1MHz. The temperature dependence of dielectric properties was observed (Figure 1). The peaks of dielectric constants were broad and occurred at progressively high temperatures as the measurement frequency is increased. This phenomenon is typical for ferroelectric-relaxor behavior.

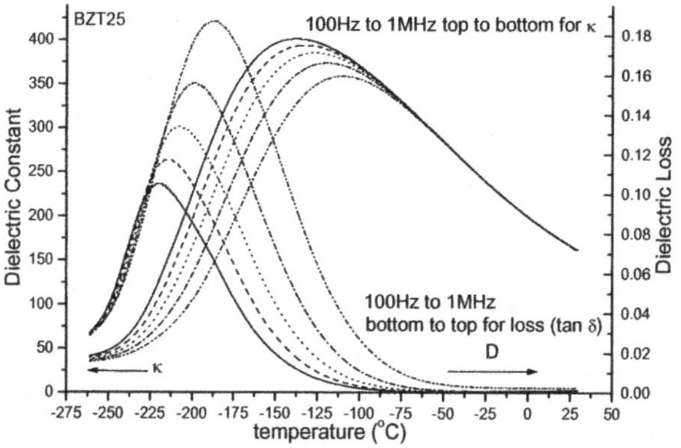

Figure 1(a). Temperature dependence of dielectric constant and loss for BZT25:MgO at 100Hz to 1MHz.

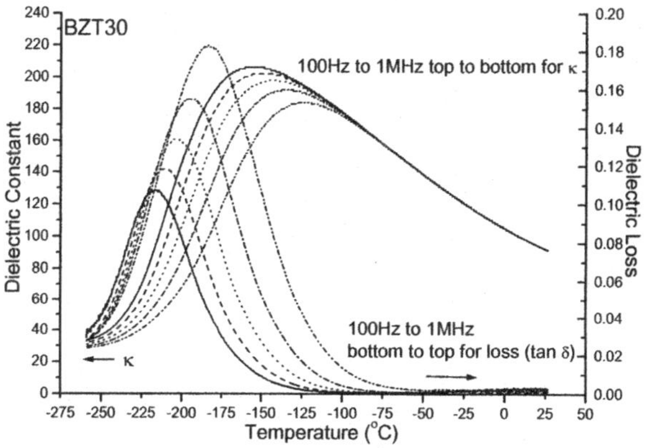

Figure.1(b). BZT30:MgO frequency and temperature dependence.

By increasing the amount of Zr that is substituted at the Ti sites, the ferroelectric-paraelectric phase transition is shifted to lower temperatures. At 10kHz, the dielectric constants for BZT25:MgO and BZT30:MgO reached a high of 374 at –120°C and 192 at –134°C, respectively. Frequency dispersion can be seen, but above –125°C, dielectric constant of BZT30:MgO is nearly independent of frequency. This also occurs for BZT25:MgO, at above –100°C.

With increasing Zr content, it can be seen that the loss peak is moved towards higher temperatures, and the transition point for permittivity is moved toward lower temperatures. The loss peaks for BZT25:MgO and BZT30:MgO were at –199°C and –195°C, respectively.

2. Dielectric Behavior with An Applied Electric Field

The temperature was held steady at specified temperatures, and the electric field was manipulated, 1) increasing from 0 to 30kV/cm, 2) decreasing to 0kV, 3) switching polarity and decreasing to –30kV/cm, increasing to 0kV, and back up to 30kV/cm. The dependence of dielectric properties (at 10 kHz) on electric field at selected temperatures for both BZT25:MgO and BZT30:MgO composites are shown in Figure 2.

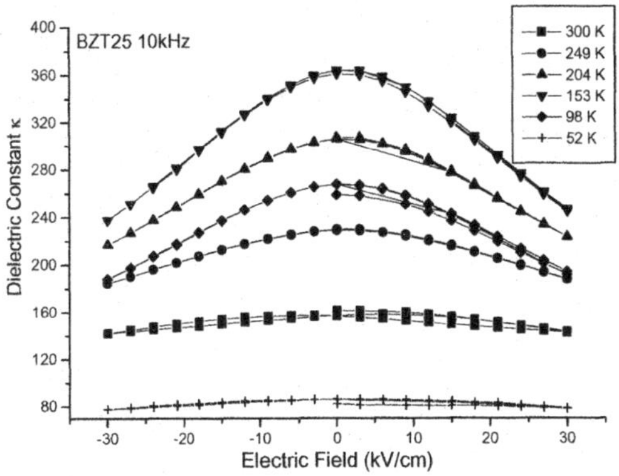

Figure 2 (a).

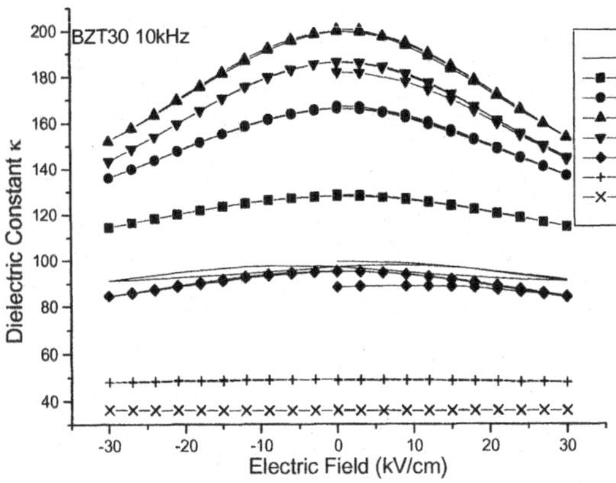

Figure 2 (b).

Figure 2. Dielectric permittivity at 10kHz as functions of bias electric field for (a) BZT25:MgO and (b) BZT30:MgO at discrete temperatures.

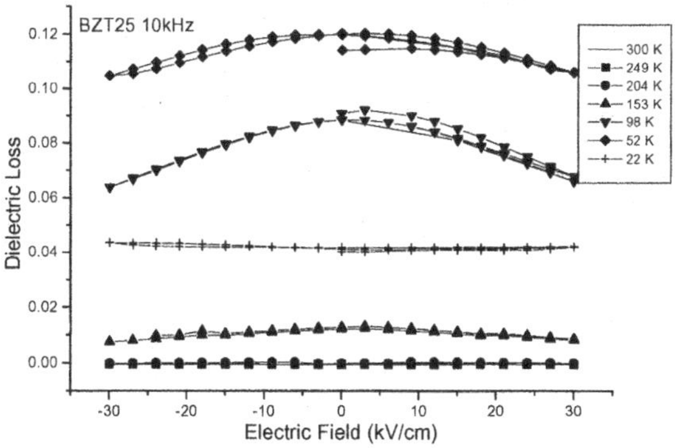

Figure 2(c)

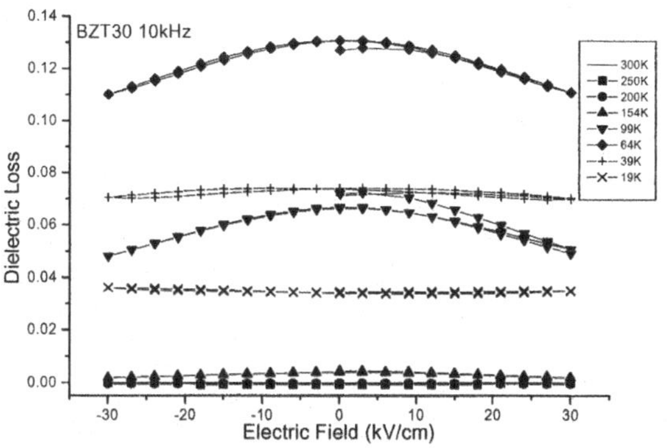

Figure 2 (d)

Figure 2. (cont.) Dielectric loss tangent at 10kHz as functions of bias electric field at for (c) BZT25:MgO and (d) BZT30:MgO at discrete temperatures.

In general the dielectric constant and loss values decrease with increasing field magnitude. The effect of electric field is noticeably stronger near the phase transition temperature, in the -175°C to -50°C temperature range (100K to 220K). The tunability is highest between -120°C and –145°C, and reaches 32.3% and 24.2% for BZT25:MgO and BZT30:MgO, respectively under 30kV/cm bias condition (see Fig. 3).

Dielectric permittivity and loss as function of temperature at 10kHz for BZT25:MgO and BZT30:MgO under discrete bias electric field are shown in Figure 4.

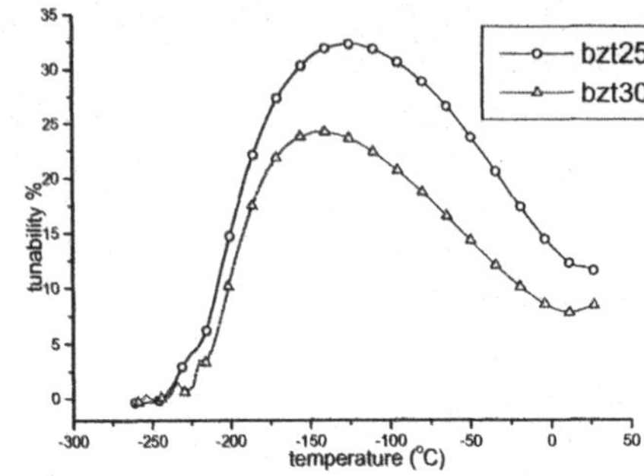

Figure 3. Tunability comparison at 100kHz.

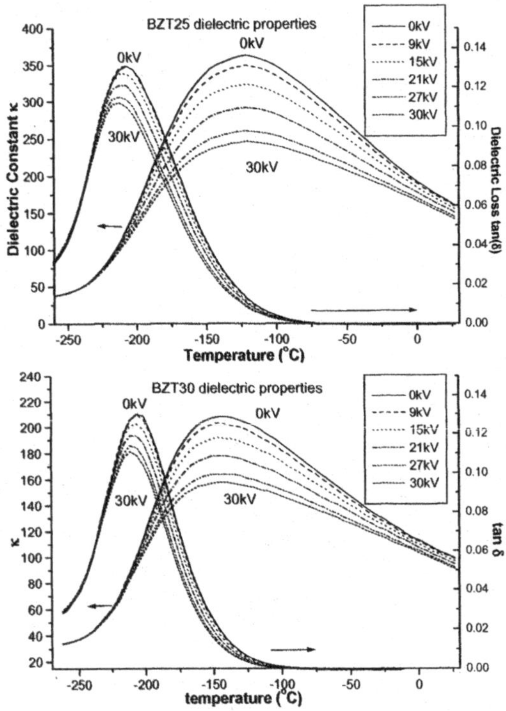

Figure 4. Dielectric constant and dielectric loss under different electric fields at 10kHz for (upper) BZT25:MgO and (lower) BZT30:MgO.

The electric field dependent dielectric properties of BZT30:MgO are similar to that of BZT25:MgO under bias electric field. Losses of BZT30:MgO are lower so is its tunability (Fig. 5). The calculated K-factor (figure of merit) is plotted in figure 6. Note dielectric loss $\tan\delta=0.0005$ at $-80^{\circ}C$ was used for the evaluation of K-factor of BZT25:MgO; dielectric loss tanδ=0.0005 at –67°C was used for the evaluation of K-factor of BZT30:MgO. The K-factor was found to reach maximum values of 645 and 484 for BZT25:MgO and BZT30:MgO respectively. Comparing to pure BZT ceramics (K-factor~ 230[5]), the maximum K-factor values obtained in the BZT:MgO composites were more than doubled.

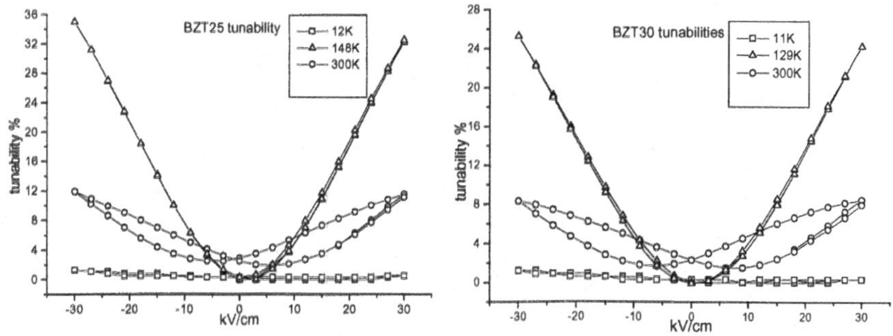

Fig. 5. Tunability dependence on field and temperature at 10kHz for (left) BZT25:MgO and (right) BZT30:MgO.

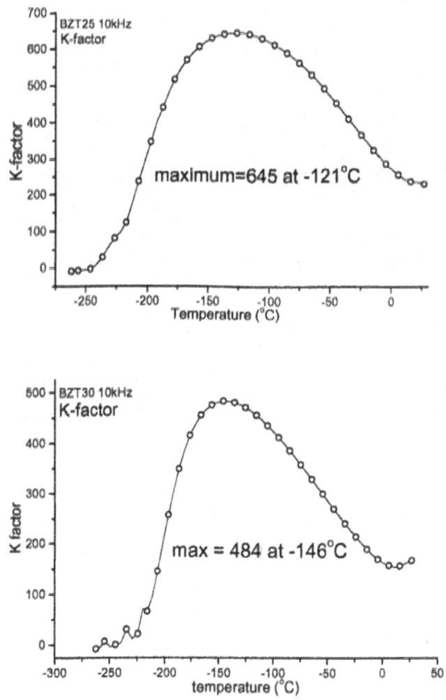

Figure 6. Temperature dependence of the K-factor for BZT25:MgO and BZT30:MgO composites.

Scanning electron microscope (SEM) study and the composition profile analysis by energy dispersive spectroscopy (EDS) showed that the MgO was randomly dispersed in the BZT matrix (see Figure 7). The tendency of MgO to absorb water has been observed in other composite ceramics and is known to impact on the conductivity of the composite material. However, water absorption did not seem to be a problem in the ceramic processing and the measurements as minimal leakage currents were observed during the high-field dielectric measurements.

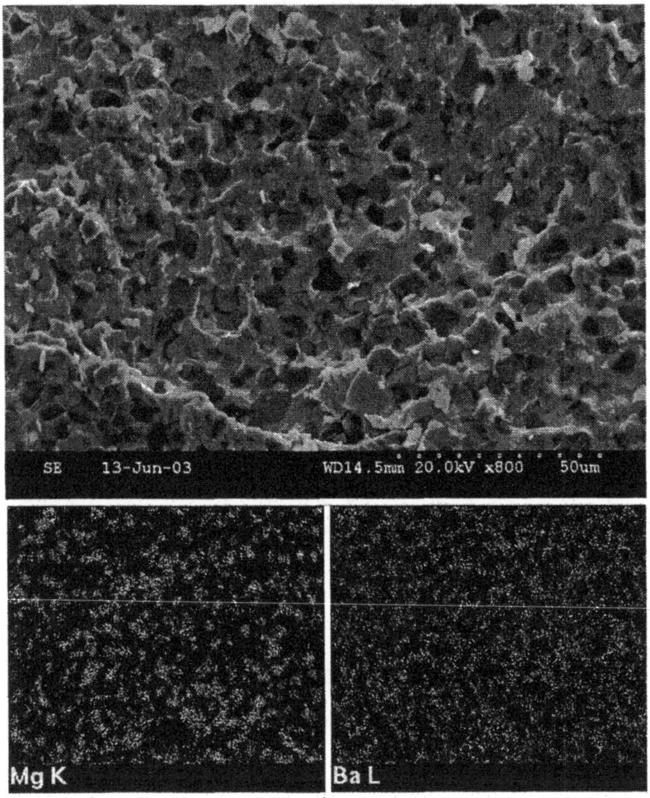

Figure 7. Microstructure of a BZT25:MgO composite: (Top) SEM image (800X magnification), (bottom) Distribution of Mg and Ba as determined by EDS (energy dispersive spectroscopy).

CONCLUSION

The microstructure and electric field dependent dielectric properties of BZT25:MgO and BZT30:MgO composites have been studied in cryogenic to room temperature range. The field dependence of dielectric constant yields high tunability, characteristic of BZT, and decreasing dielectric constant attributing to MgO additive. The dielectric constants are tailored down to suitable values for microwave applications whereas the figure-of-merit K-factor is either comparable or significantly improved, depending on desired application temperature, in comparison to the pure BZT. The results further show that BZT-based materials are promising candidates that can be used in applications ranging from tunable filters to antennas. As a lead-free material, BZT is environmentally friendly, and can be useful in current attempts to reduce usage of lead containing piezoelectric materials.

In conclusion, BZT25:MgO and BZT30:MgO composite ceramics show suitable dielectric constant value (hundreds instead of thousands), high tunability (32.3% and 24.2% for BZT25:MgO and BZT30:MgO, respectively under 30kV/cm bias) in the -145°C to -120°C range, low losses (lower than 0.001 at temperatures above -50°C), and high K-factor (higher than 600 for BZT25:MgO at ~-120°C)). By varying the Zr:Ti ratio and optimizing the percentage of MgO phase in $Ba(Zr_xTi_{1-x})O_3$-MgO composite, one would-be able to optimize the property for specific application at desired temperature.

ACKNOWLEDGEMENT

This material is based upon work supported by the National Science Foundation under Grant No. EEC-0244030.

REFERENCES

[1] D. Galt, J. Price, J.A. Beall, and R.H. Ono, "Characterization of a tunable thin film microwave $YBa_2Cu_3O_{7-x}/SrTiO_3$ coplanar capacitor," *Appl. Phys. Lett.* **63**, 3078 (1993)

[2] P.W. Rherig, S.E. Park, S.T. McKinstry, G.L. Messing, B. Jones, and T.M. Shrout, "Piezoelectric properties of zirconium-doped barium titanate single crystals grown by templated grain growth," *J. Appl. Phys.* **86**, 1657 (1999).

[3] Yu Zhi, R. Guo, A. S. Bhalla, "Orientation Dependence of the Ferroelectric and Piezoelectric Behavior of $Ba(Ti_{1-x}Zr_x)O_3$ Single Crystals, "*Appl. Phys. Lett.*, **77**(10), 1535-1537, (2000).

[4] Yu Zhi, Ang Chen, Ruyan Guo, and A.S. Bhalla, "Ferroelectric-Relaxor Behavior of $Ba(Ti_{0.7}Zr_{0.3})O_3$," *J. Appl. Phys.*, **92**(5), 2655-2657 (2002).

[5] Yu Zhi, Ang Chen, Ruyan Guo and A.S. Bhalla, "Dielectric Properties and High Tunability of $Ba(Ti_{0.7}Zr_{0.3})O_3$ Ceramics under dc Electric Field," *Appl. Phys. Lett.*, **81(7),** 1285-1287 (2002).

[6] S. Agrawal, H. Manuspiya, R. Guo, D. Agrawal and A.S. Bhalla , "Dielectric Tunability of Microwave Sintered BST:MgO Composites" *Ceramic Materials and Multilayer Electronic Devices*, Ceramic Transactions, V. 150 (2004).

[7] E.F. Alberta, R. Guo and A.S. Bhalla, "Novel BST:$MgTiO_3$ Composites for Frequency Agile Applications," *Ferroelectrics,* **268,** 169-174 (2002).

Processing, Texture and Grain Boundaries in Perovskites

MICROSTRUCTURAL ASPECTS OF INCLINED SUBSTRATE DEPOSITION TEMPLATES FOR COATED CONDUCTORS

Dean J. Miller[1], Susana Trasobares[1], Rachel Koritala[2],
Beihai Ma[2], Krishna Uprety[2], Uthmalingam (Balu) Balachandran[2]
[1]Materials Science Division
[2]Energy Technology Division
Argonne National Laboratory
Argonne, Illinois 60439

ABSTRACT:

Inclined substrate deposition (ISD) has been developed as an efficient and economical method to produce biaxially textured templates useful in the fabrication of $YBa_2Cu_3O_x$ (YBCO) coated conductors. In this work we have focused on an architecture that consists of an ISD MgO layer capped with a thin homoeptiaxial layer of MgO and a single perovskite ($SrRuO_3$) buffer layer. The single perovskite buffer layer promotes improved texture and eliminates the necessity for deposition of additional layers that reduce the overall efficiency of the ISD approach. With a terminal surface consisting of a perovskite, epitaxial growth of YBCO leads to an inclined orientation. Compared to ISD-based conductors prepared using a CeO_2 buffer, this architecture typically exhibits better transport critical current densities that can be correlated with an improvement in texture attributed to the perovskite buffer.

INTRODUCTION:

Significant progress has been made in the development of practical conductors based on high temperature superconductors. The advances made in "first generation" wires based on the Bi-Sr-Ca-Cu-O superconductors have now been followed by similar progress in "second generation" wires. These $YBa_2Cu_3O_x$ (YBCO) coated conductors offer significant promise for a variety of high-current and electric power applications.[1,2,3] A critical aspect of coated conductors is a strong biaxial texture that mitigates the weak-link effect associated with high angle grain boundaries in high temperature superconductors.[4] The goal is to produce a high degree of biaxial texture in the superconducting layer, and in most approaches this is achieved by epitaxial growth of the superconducting layer (YBCO) onto a suitably textured template.

A variety of techniques have been pursued to produce suitably textured templates, with three techniques attracting the most attention: ion-beam-assisted deposition (IBAD), rolling-assisted biaxially textured substrates (RABiTS), and oblique or inclined substrate deposition (ISD).[5,6,7] Each of these approaches pose different advantages and challenges. A common aspect of each approach is the need to achieve a high degree of texture in the YBCO layer based on the texture in the underlying template. In this work, we have focused on an ISD-based architecture that utilizes a single perovskite ($SrRuO_3$) buffer layer. The single perovskite buffer layer promotes improved texture and eliminates the necessity for deposition of additional layers that reduce the overall efficiency of the ISD approach. An increase in transport critical current density can be correlated with the improvement in texture over alternative ISD architectures.

EXPERIMENT:

The samples studied in this work are all based on an underlying ISD-MgO layer. The ISD layer is deposited at room temperature by electron-beam evaporation from a metallic Mg source. In some cases, a second MgO layer is deposited on top of the ISD layer. This deposition is carried out by e-beam evaporation, as with the ISD growth, but at high temperature, typically 700°C, yielding a homoepitaxial MgO layer. Subsequent deposition of the perovskite buffer layer ($SrRuO_3$) and the final $YBa_2Cu_3O_x$ (YBCO) layer is carried out by pulsed laser deposition. Details of each of these processes have been reported previously.[7,8]

The primary characterization methods used in this work included x-ray diffraction/pole figures, scanning electron micrsocopy (SEM), and transmission electron microscopy (TEM), each using standard commercial instrumentation. Cross-sectional specimens were prepared for TEM by conventional methods involving sectioning, grinding and polishing, followed by ion beam thinning to final perforation. Microscopy was carried out in the Electron Microscopy Center at Argonne.

RESULTS AND DISCUSSION:

Generally, ISD-based conductors consist of a metallic substrate (untextured), the template layers that establish biaxial texture (based on ISD MgO), buffer layers, and the YBCO layer (with a stabilization layer on top). Often, the template layer consists of an ISD layer with an additional homoeptiaxial layer of MgO deposited on top of it. The critical aspect for ISD-based coated conductors is the same as that for other coated conductor architectures – the goal is to produce a long-length conductor that has few or no high angle grain boundaries that will limit current flow. As shown in Figure 1, the common strategy is to produce a polycrystalline template that mimics a single crystal.

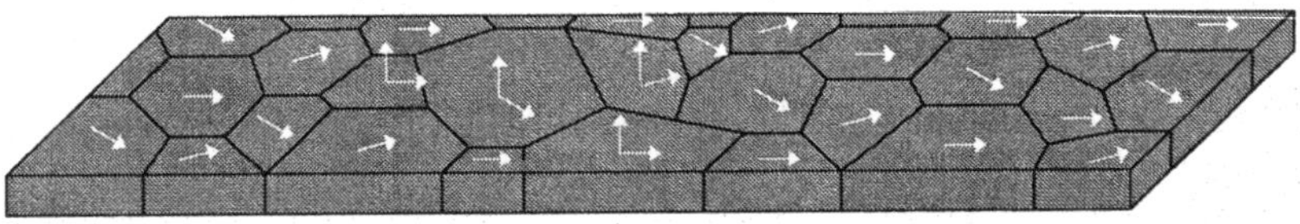

Figure 1: Schematic illustration of a biaxially texture template for coated conductor fabrication. Such templates are polycrystalline, but with only low angle grain boundaries between adjacent grains.

The mechanism for texture development during inclined substrate deposition is based on shadowing and competitive growth such that favorably oriented grains outgrow their neighbors. [9] In this mechanism, texture develops as a function of thickness. It has been shown that during ISD growth, a strong texture develops very quickly. In order to evaluate texture development, we utilized selected area electron diffraction to measure texture as a function of position in the ISD layer. In this approach, selected area diffraction patterns are collected as a function of position in the ISD film, and the texture is evaluated by the broadening of the fundamental reflections. The full-width of the reflection is used as an analogy to x-ray approaches that use peak full-width at half maximum (FWHM) as the measurement of texture, as shown in Figure 2. This experiment was carried out on several different ISD samples in orientations parallel to the inclination axis and perpendicular to that axis, all with similar results. It is important to note that quantitative measurement of texture from electron diffraction is challenging, and we have not carried out a fully quantitative analysis. For this reason, the degree of texture measured here cannot be directly compared to that from x-ray techniques, for example, or even from sample to sample using this approach. However, in each experiment we have maintained constant sample thickness, beam intensity, and exposure time so the trend and relative change in texture within a given measurement is reliable. From this data, it is seen that a strong texture develops very quickly, within about half a micron, and then improves much more slowly. Taken together with surface roughness data that shows a minimum at about the same thickness, a thickness of about half a micron for the ISD layer may be viewed as optimum.

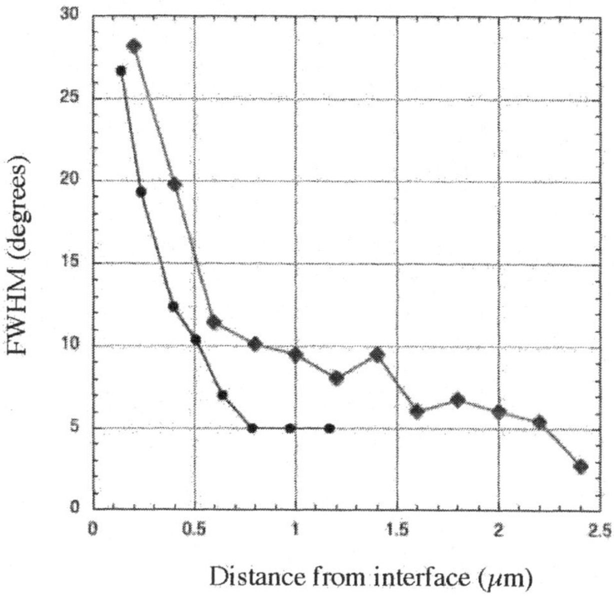

Figure 2: In-plane textures as measured by selected area diffraction as a function of distance from the substrate interface. Each data point represents the angular spread in diffraction reflections recorded from a small, selected area. See text for details.

While the ISD-MgO layer provides a good template for texture, we have found it useful to deposit an additional layer of MgO at high temperature. This layer grows homoepitaxially on the ISD MgO and provides a smoother, more dense layer upon which subsequent layers may be deposited. As an additional advantage, the homoepitaxial layer yields a modest improvement in texture over that in the ISD MgO, typically improving in-plane texture by a degree or two. The mechanism for this texture improvement has been described previously.[10] An example of this ISD-homoepitaxial architecture is shown in Figure 3. In this cross-sectional view, it can be seen that the homoeptixial layer tends to smooth the surface and is much more dense than the underlying ISD layer.

The motivation to use a single buffer layer is that it promotes the efficiency of template development that is inherent in the ISD process. The rationale for use of a perovskite as the buffer layer is based on several factors. Firstly, perovskites offer a high degree of flexibility in terms of chemical compatibility and physical properties within the same crystallographic framework. Secondly, perovskites have been demonstrated to provide a compatible buffer with MgO and, perhaps more significantly, have been shown to yield significant improvement in texture over the underlying template layer. Jia, et al., for example, demonstrated a significant improvement in texture associated with the deposition of $SrRuO_3$, $SrTiO_3$, and $LaMnO_3$ perovskite buffers.[11]

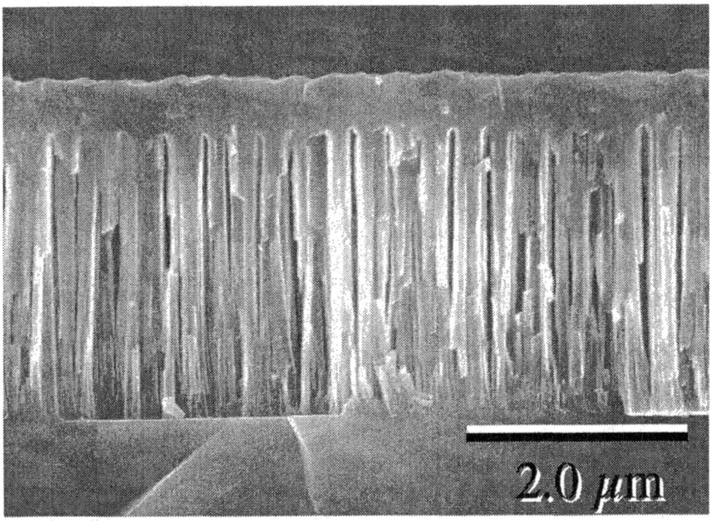

Figure 3: Cross-sectional scanning electron microscope image of the ISD architecture showing the columnar ISD MgO grains capped by the more dense homoepitaxial MgO layer at the top.

The evolution of texture from the underlying MgO layers through the $SrRuO_3$ buffer to the YBCO conductor layer is illustrated by the x-ray diffraction data shown in Figure 4. The improvement of texture as a function of continued growth is a feature that has been observed for each of the approaches to biaxially textured templates. However, the substantial improvement in texture associated with a perovskite buffer is a more significant benefit for the ISD templates studied here since they do not exhibit texture comparable to other templates (e.g. IBAD and RABiTS). Thus, after deposition of the perovskite buffer, a more substantial improvement in texture is realized. For YBCO on SRO-terminated templates the values are below 10°, with recent results in the range of 6-8°, considerably better than the underlying MgO. The buffer layers grow epitaxially with a cube-on-cube orientation relationship with the underlying MgO, meaning the buffer layers also are in the "inclined" orientation.

The YBCO deposited onto perovskite-terminated templates also grows with a cube-on-cube orientation, meaning the YBCO is inclined as are all the underlying layers. For deposition onto SRO, the lattice matching with YBCO is almost perfect since the lattice constants and sublattice structure are essentially identical. Therefore, even for "inclined" growth, *a-b* planes can be well-terminated and cube-on-cube epitaxial growth is favored.

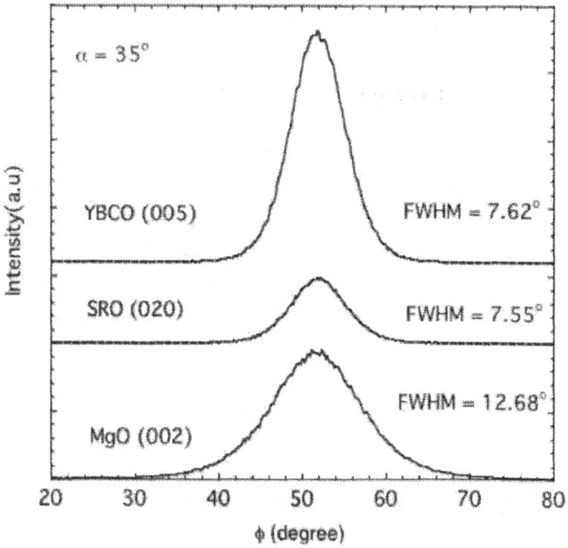

Figure 4: X-ray diffraction measurements of out-of-plane texture using the reflections indicated. There is a considerable improvement in texture associated with deposition of the perovskite (SRO) layer that yields an improvement in texture in the YBCO conductor layer.

Figures 5, 6 and 7 show examples of the buffer layer structure and the YBCO in cross-section. Figure 5 is a cross-section imaged by SEM. This image of a reasonably thick film, approximately 1.6 μm, shows the overall structure and illustrates the terraced surface associated with inclined growth. Transport measurements on this sample were over 1 mA/cm^2, an encouraging value for YBCO films of this thickness.

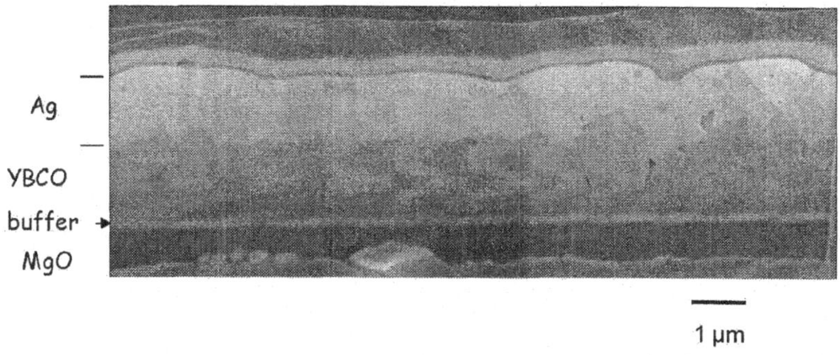

Figure 5: SEM cross-sectional image showing the overall architecture of these ISD-based conductors. The YBCO film is about 1.6 μm thick and exhibits a terraced surface associated with the inclined orientation acquired from the underlying template layers.

Figure 6 shows the buffer layer structure in cross-sectional TEM. The SRO buffer exhibits an epitaxial orientation relationship with the underlying MgO, and the YBCO grown on top of it also exhibits an epitaxial orientation. Figure 7 shows the YBCO layer in more detail, and the inclined nature of the YBCO is indicated by the c-axis fringes and stacking faults evident in the image. The high density of stacking faults suggest that further improvement in the deposition of YBCO on this buffer can be realized.

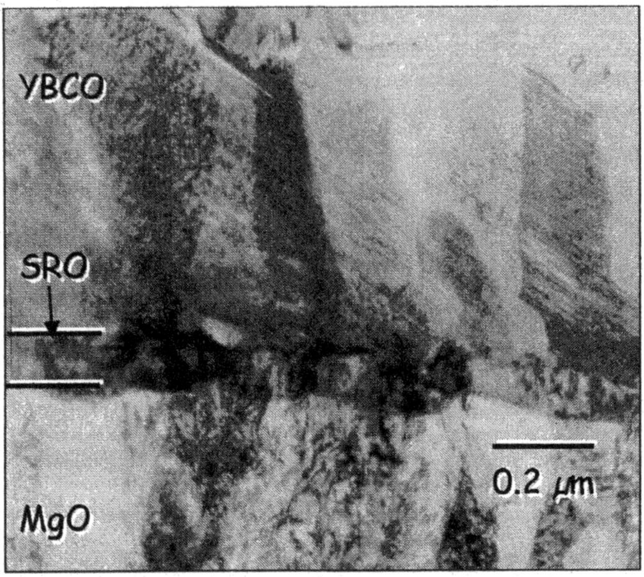

Figure 6: TEM cross-sectional image of the $SrRuO_3$ buffer layer.

CONCLUSIONS:

In this work we have explored a simplified architecture based on an ISD template. In this architecture, a single perovskite buffer layer is used as the terminal layer of the template, upon which the YBCO layer is deposited. These results provide encouraging evidence that ISD can be a viable alternative to other approaches for coated conductors. The ISD process itself yields a template layer with a high degree of biaxial texture, with an efficient development of good texture at high speed that is independent of substrate and has demonstrated scale-up potential. Perovskite buffer layers grow in an epitaxial, cube-on-cube orientation that preserves the "inclined" orientation of the underlying MgO and, significantly, provide a further improvement in template texture. Deposition of the YBCO conductor onto this architecture yields conductors with good texture that exhibit reasonable critical current densities. Further optimization of both the buffer layer and conductor deposition for this specific architecture may be expected to yield additional improvement in properties.

ACKNOWLEDGEMENTS:

This work was supported by the U.S. Department of Energy, Offices of Science and of Electric Transmission and Distribution, under contract #W-31-109-Eng-38. Electron microscopy was carried out in the Electron Microscopy Center at Argonne National Laboratory, which is supported by the Office of Science.

Figure 7: TEM cross-sectional image of the YBCO layer deposited on top of the $SrRuO_3$ buffer. The inclined orientation is evident by the c-axis fringes and c-axis stacking faults. These faults suggest further improvement in YBCO deposition on these buffers may lead to better film quality.

REFERENCES:

1 D.K. Finnemore, et al., *Physica C* 320, 1-8 (1999).
2 Y. Iijima, K. Matsumoto, *Supercond. Sci. & Technol.* 13, 68-81 (2000).
3 J.O. Willis, et al., *Physica C* 335, 73 (2000).
4 D. Dimos, P. Chaudari, J. Mannhart, *Phys. Rev.* B 41, 4038-4049 (1990).
5 Y. Iijima, M. Kimura, T. Saitoh, K. Takeda, Physica C 355, 15 (2000).
6 A. Goyal, et al. Appl. Phys. Lett. 69, 1975-1977 (1996).
7 B. Ma, et al., Physica C 366, 270-276 (2002).
8 B. Ma, et al., Supercond. Sci. & Technol. 16, 464-472 (2003).
9 Paritosh, D.J. Srolovitz, J. Appl. Phys. 91 (4) (2002)
10 D.J. Miller, et al., Proc. Electrochemical Soc.Symp. L1. Orlando, FL, Oct. 12-17 (2003)
11 Q.X.Jia, et al., Appl. Phys. Lett. 81 (24) 4571 (2002)

TWINS IN SUPERCONDUCTING MELT-TEXTURED GROWN Y-Ba-Cu-O AS RELATED TO CRITICAL CURRENT DENSITIES

Siu-Wai Chan, and Linfeng Mei
Columbia University,
Department of Applied Physics and Applied Mathematics,
School of Engineering and Applied Science,
New York, NY 10027.

ABSTRACT

Twin spacings in melt-textured grown (MTG) Y-Ba-Cu-O (YBCO) are refined by annealing at higher temperatures. This isothermal method exploits the temperature dependence of twin-boundary-energy. The twin boundary energy is obtained by two independent methods: (1) twin spacing and (2) twin-tip shape, employing transmission electron microscopy. We observe that twin spacing decreases from 190 nm to 54 nm as the temperature increases from 450 to 680°C. Critical current measurements support that a sample with a higher density twin density often results in a higher pinning-force i.e. 3.4 x 10^8 N/m^3 versa 1.2 x 10^8 N/m^3.

INTRODUCTION

Lamellae twin structure in superconducting $YBa_2Cu_3O_{7-x}$ (123) is ubiquitous originated from the shear strain during the tetragonal-to-orthorhombic transformation. The associated twin boundaries (TBs) are flux-pinning centers in intermediate temperatures. The early controversy was caused by the anisotropic nature of TBs in flux-pinning[1]. Specifically, they act as easy conduits for vortices traveling along the boundaries but effectively pin transverse fluxes. Refinement of twin microstructure in melt-textured $YBa_2Cu_3O_{7-x}$ can be engineered by a number of different methods. For examples: (1) doping with PtO_2 [2], (2) adding an optimal volume fraction of finely dispersed Y_2BaCuO_5 (211) particle inclusions [3], and (3) annealing at different temperatures. Here we describe the third method, the isothermal method, can be accomplished with systematic microstructure study.

EXPERIMENTAL

Melt textured grown (MTG) YBCO pellets with initial composition of 28.6wt% 211 and 71.4 wt% 123 and 0.5wt% PtO2 are prepared as described elsewhere [3]. Square pieces of 3 x3 x1.0-1.3mm are or cut from the MTG pellets and isothermally annealed at 450, 500, 600, 650 and 680°C. The superconducting properties of the samples are then measured by a superconducting quantum interference device (SQUID) magnetometer (made by Quantum Design) with the magnetic field (H) applied in the [001] direction. All measurements are performed at 77K. The critical current density Jc as a function of magnetic field can be calculated from the Beam critical state model [4]. The pinning force is then calculated from the cross-product of the critical current density and the magnetic field. Samples are then prepared for the electron transparency and examined with a JEOL 100CX transmission electron microscope.

RESULTS AND DISCUSSION

The microstructure of MTG pellets is a matrix of 123 with dispersed 211 particles. The 123 matrix is essential a single grain of $YBa_2Cu_3O_{7-x}$ interspersed with 211 particles. Our focus is on the twin microstructure of the 123 matrix. Therefore, 211 particles are not shown here. Two sets of electron micrographs are obtained. One set gives information on average twin spacing as shown in Fig. 1. Figure 1 is a transmission electron micrograph showing the twin spacing (Tw) within a sample that has been annealed at 600°C. Many micrographs are obtained to give statistically meaningful twin spacing and twin domain size (G) for each sample.

The other set is obtained to provide information for the shape of twin tips from samples that are arrested from full annealing (see Fig. 2). Basically, the Gibbs free energy of twin boundary energy can be given by special parameter fitting of tip shape employing dislocation pile-up model with a thin twin assumption [2]. Provided the entropy term is positive, generally an increase in temperature gives lower boundary energy. Exploiting lower twin boundary energy with higher annealing temperature, we prepare the MTG samples with progressive finer twins in the 123 matrix.

The pinning forces (Fp) in samples annealed at different temperatures as a function of magnetic field is shown in Fig. 3. Each pinning force is calculated from the cross-product of the measured critical current density and the applied magnetic field. There is a progressive increase of the pinning force with a higher annealing temperature as well as a clear maximum pinning force for the sample annealed at 680°C. Table I gives a summary of the average twin spacing, maximum pinning force of each samples and other related properties. There is a correlation between the finer twin spacing with the increase temperature and the increase pinning force. Figure 4 shows the field at maximum pinning force increases with annealing temperature.

CONCLUSIONS

We employ the temperature dependence of the twin boundary energy to refine twin structure and increase Jc in $YBa_2Cu_3O_{7-x}$. The detailed measurements of twin boundary energy using the twin spacing and twin method will be described elsewhere [5]. The critical current density, the pinning force and the field of maximum pinning all increase with increasing annealing temperature. This isothermal method provides a viable way to improve these superconducting properties that are extrinsic and a strong function of the microstructure. Twin spacing reduces from 190 nm to 54 nm as the annealing temperature increases from 450 to 680°C. Critical current measurements confirm that a higher density twin density often results in a higher pinning-force i.e. 3.4×10^8 N/m^3 versa 1.2×10^8 N/m^3.

ACKNOWLEDGMENTS

This work is supported by National Science Foundation.

Table I. Property summaries of the isothermal samples: annealing temperatures, critical current density at 77K of magnetic field H { $J_c(H,77K)$}, maximum flux-pinning force $F_{p,max}$, irreversible field $H_{irre.}$, the magnetic field at which F_p is largest $H_{max.}$, estimated twin domain size G, twin spacing range and average twin spacing (T_w), twin density.

Sample	450	500	600	650	680
Annealing temperature	450 ^{0}C	500 ^{0}C	600 ^{0}C	650 ^{0}C	680 ^{0}C
J_c(0.75Tesla) (A/cm^2)	15600	14000	18700	16900	35000
J_c (1Tesla) (A/cm^2)	9800	12500	17270	15230	31920
J_c(1.25Tesla) (A/cm^2)	5600	10700	15000	13400	26600
J_c(1.5Tesla) (A/cm^2)	2900	8600	12100	11400	21300
$F_{p,max.}$ (10^8 N/m^3)	1.17	1.34	1.87	1.71	3.35
$H_{max.}$ (Tesla) *	0.75	1.25	1.25	1.50	1.5
$H_{irre.}$ (Tesla)	3.00	4.25	3.75	3.75	4.5
Estimated grain size G by TEM (μm)	35,30,25, 22.5,53	25, 31, 45	28, 22.5 15	24, 32	18, 11.5
T_w range (Å) (TEM)	1000 / 2500	990 / 1710	784 / 1716	965 /1330	460/ 623
Ave. T_w (Å) (TEM)	1900	1218	1171	1134	539
Twin boundary density (1/μm)	5.26	8.21	8.54	8.82	18.55

REFERENCES

1. L. Ya. Vinnikov, L. A. Gurevich, G. A. Yamelchenko and Yu. A. Ossipyan, "Direct observation of the lattice of Abrikosov vortices in high-T_c superconductor $YBa_2Cu_3O_x$ single crystals", *Solid State Commun.*, **67**, 421 (1988); C. A. Duran, P. L. Gammel, D. J. Bishop, J. P. Rice and D. M. Ginsberg, "Role of twin boundaries in the magnetic flux penetration in $YBa_2Cu_3O_{7-\delta}$", *Phys. Rev. Lett.*, **74**(18), 3712 (1995); U. Welp, T. Gardiner, D. O. Gunter, B. W. Veal, G. W. Crabtree, V. K. Vlasko-Vlasov and V. I. Nikitenko, "Welp et al reply", *Phy. Rev. Lett.*, **74**(18), 3713 (1995); U. Welp, T. Gardiner, D. Gunter, J. Fredrich, G. W. Crabtree, V. K. Vlasko-Vlasov and V. I. Nikitenko, "Magneto-optical study of twin boundary pinning in $YBa_2Cu_3O_{7-\delta}$", *Physica C*, **235-240**, 241-244 (1994); and I. Maggio-Aprile, Ch. Renner, A. Erb, E. Walker and Ø. Fischer, "Direct Vortex Lattice Imaging and Tunneling Spectroscopy of Flux Lines on $YBa_2Cu_3O_{7-\delta}$", *Phy. Rev. Lett.*, **75** (14), 2754 (1995).
2. V. S. Boyko, Siu-Wai Chan and Manoj Chopra, "The shape of a twin as related to the elastic forces acting on twinning dislocations in $YBa_2Cu_3O_{7-\delta}$", *Phys. Rev. B*, 63 224521(2001).
3. Manoj Chopra, Siu-Wai Chan, R. L. Meng and C. W. Chu, "Y_2BaCuO_5 addition and its effects on critical currents in large grains of $YBa_2Cu_3O_{7-\delta}$: A quantitative microstructural study", *J. Mater. Res.*, **11**, 1616 (1996); and Oratai Jonprateep, and Siu-Wai Chan, "Twin Spacing and Its Correlation with Critical Current Density in Melt Textured YBCO with Yttria Nanoparticle Addition", IEEE Transactions on Applied Superconductivity, **13** p. 3502, 2003.
4. C.P. Bean, "Magnetization of hard superconductors", *Phys. Rev. Lett.*, **8**, 250(1962)
5. Linfeng Mei and Siu-Wai Chan to be published.

Fig.1 TEM micrograph showing parallel {110} twin boundaries viewed along [001] direction in a sample annealed at 600C.

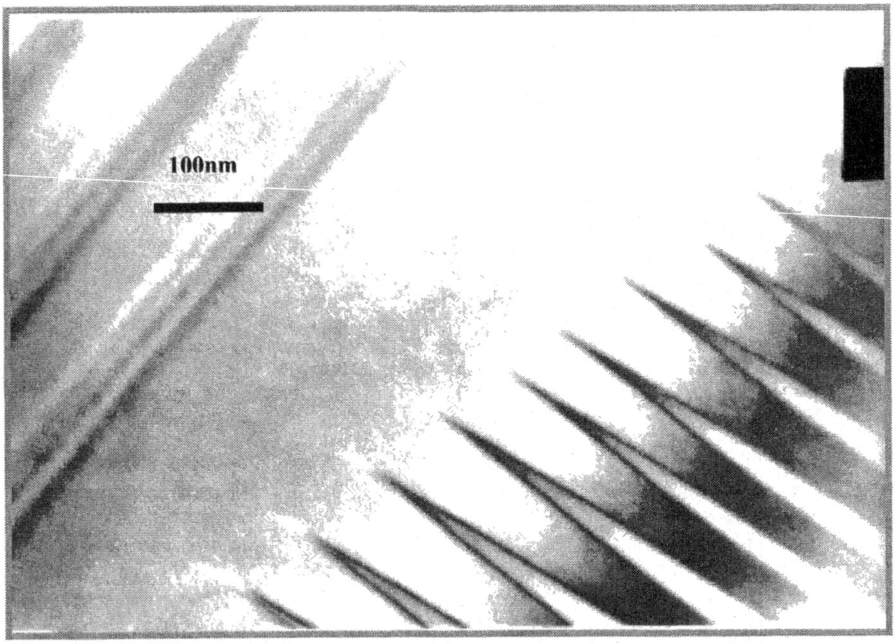

Fig. 2 Advancing twin tips of the (110) variant in a sample after arrested anneal at 650°C. See the parallel twin boundaries of the $(1\bar{1}0)$ variant at the upper-left corner.

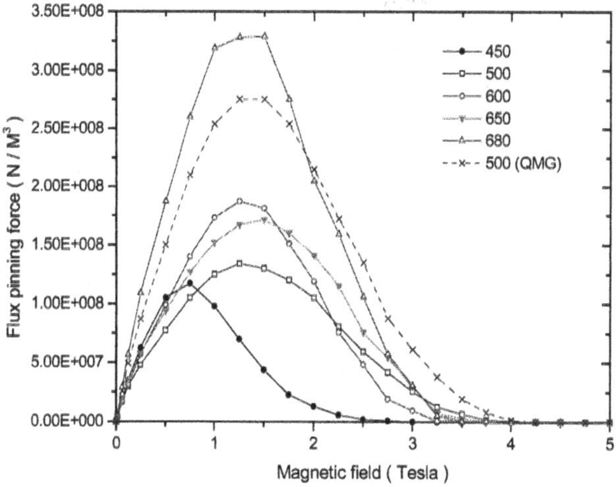

Fig. 3 The field dependence of the flux pinning forces of the isothermally annealed samples

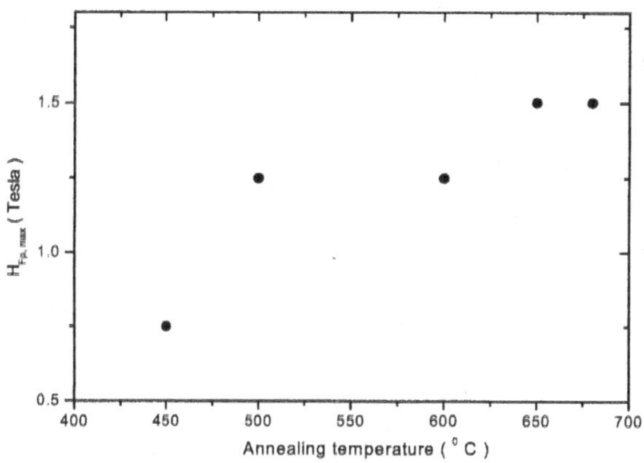

Fig. 4. Magnetic field at maximum pinning force ($H_{Fp.max}$). – Annealing temperature correlation for isothermal samples

CHARACTERIZATION OF LANTHANUM CHROMITE POWDERS OBTAINED BY COMBUSTION REACTION

Adolfo Franco Júnior* and Cirilo Gomes Valente Neto
Núcleo de Pesquisa em Química,
Universidade Católica de Goiás,
C.P. 86, Setor Universitário, Goiânia-GO, Brazil,
email: franco @ucg.br

ABSTRACT

Fine particle lanthanum chromite has been prepared by the combustion of aqueous solutions containing lanthanum nitrite, chromium (III) nitrate and urea in two stoichiometric amounts: Fuel (urea) stoichiometry, batch #1, and an excess fuel (urea), batch #2. The solutions, when heated up to 300°C, ignited and yielding gray-wish powder of crystalline $LaCrO_3$. Powders were characterized by X-ray diffraction. The particle-size of powders produced by an excess of fuel (urea) were slightly smaller than those of fuel stoichiometry. The fine and also uniform particle size distribution ensured their sinterability. Pellets were sintered in air at 1600°C for 2 h achieved 95% of the theoretical density.

INTRODUCTION

Lanthanum chromite ($LaCrO_3$) based perovskites is one of the most promising material for application in the solid oxide fuel cell (SOFC) as an interconnecting material due to its high electrical conductivity >100 $\Omega^{-1}m^{-1}$ above 300°C, and chemical stability.[1-2] It is very difficult to sinter $LaCrO_3$ due to the high vapour pressure of the chromium component.[3-4] At very high temperatures in strongly oxidizing atmosphere, the material oxidizes, producing a volatile product, and this process can occur at a significant rate in air above 1600°C. Temperature as high as 1775°C and oxygen activities in the range from 10^{-12} to 10^{-19} atm have been required for the densification of the material.[3] Thus it leads the study of chemical methods of synthesis. Pechini's method and combustion reaction method are the two techniques which use liquid solutions of the desired metal ions to form the respective oxide powder.

Pechini's method has its basic principle the chelation of complex cations forming a soft mixture and porous resin intermediate stage.[5] The obtained mixture is then pulverized and calcined.

The combustion reaction method has been largely studied by Patil[6] to produce a variety of powders. This method involves the combustion of redox mixtures containing the desired metal ions in oxidizing reactants (O) and the fuel (F) as a reducing reactant, where, in accordance with the thermochemical concepts, used in propellant chemistry,[7] the total oxidizing and reducting potentials are stoichiometrically balanced to release the maximum energy for the reaction.[6] According to the concepts used in propellant,[7] the elements C, H, La, Cr, are considered as reducing elements with the corresponding valencies+2, +1, +3 and +3, respectively. The element oxygen is considered as an oxidizing with the valency -2. The valency of nitrogen is considered to be zero. Thus, based on oxidizing and reducing valencies of the compounds used in the combustion mixtures can be easily calculated.

* Member, American Ceramic Society

The present study describes the combustion synthesis of fine uniform and pure particles of $LaCrO_3$ perovskite-type powders using metal nitrite (oxidizer) and urea (fuel) mixtures ignited at 300°C, under normal atmospheric pressure. The role of the fuel stoichiometric on the composition behavior as well as its effect on the particle characteristics have been investigated too.

EXPERIMENTAL PROCEDURE

Fine particle $LaCrO_3$ has been prepared by rapidly heating an aqueous saturated solution containing stoichiometric amounts of Lanthanum nitrite, chromium nitrite and urea mixture.

Two batches were prepared according to the stoichiometric composition. The combustion mixture for the preparation of $LaCrO_3$ contains La, Cr and urea in appropriate molar ratio. In the first batch, named hereafter as batch #1, the molar ratio was 1:1:5 of $La(NO_3)_3.6H_2O$ (Merk, Darmstadt, FRG), $Cr(NO_3)_3.6H_2O$ (Merk, Darmstadt, FRG) and $CO(NH_2)_2$ (Merk, Darmstadt, FRG), respectively which correspond to stoichiometric reaction. In the second batch, named hereafter as batch #2, the molar ratio was 1:1:10 of $La(NO_3)_3.6H_2O$, $Cr(NO_3)_3.6H_2O$ and $CO(NH_2)_2$, respectively which correspond to fuel-rich stoichiometric reaction. i.e. excess of urea. Small amount of water, c.a. 5% weight, was added to the reactants in a widemouthed vitreous silica basin of 350 cm^3 capacity and the resulting solution was heated up on a hot plate maintaining at 300°C inside a fume cupboard, under ventilation. Initially, the solution boils and undergoes dehydration followed by decomposition generating combustible gases such as oxides of nitrogen, HNCO and NH_3.[8] These gases burnt with an appearance of a flame yielding a foamy, loosely bound particles and dry powder.

Room temperature X-ray diffraction (XRD) pattern of as prepared powder was recorded using a Shimadzu diffractometer (model 6000) with CuK_α (λ=1.5406 Å) in a wide range of Bragg angles ($5^\circ \leq 2\theta \leq 80^\circ$) with a scanning rate of 2°/min.

Powder densities were determined by Archimedes method in isopropyl alcohol and sedimentation method (Model SKC 2000 photosizer, Micron) to ascertain the particle size, particle shape, and the extent of agglomeration in the powders.

Pellet samples in cylindrical shape (8 mm diameter and 3 mm thickness) of powders from both batches were cold isostatically pressed at 250 MPa. Polyvinyl alcohol (PVA) was used as binder to reduce the brittleness of the pellets. The binder was burnt out during the sintering process. The pellets were sintered in an alumina crucible at 1600°C in a reducing atmosphere for two hours. Pellets densities were determined by Archimedes method.

RESULTS AND DISCUSSION

Using the stoichiometric molar composition of mixtures, lanthanum chromite powders were prepared. Fig. 1 (a) shows a green-wish liquid solution, before ignition, followed by the flame produced by the ignition of the gases during the combustion reaction, Fig. 1 (b). The combustion reaction is extremely exothermic and the system can reach, in few minutes, temperatures as high as the sintering temperature of $LaCrO_3$.[6] After all the gases have been burnt out, is yielded a foamy, loosely bound particles and dry grayish powder, Fig 1 (c). It was observed that the fuel-rich solution ignited easily, the high temperature was sustained longer and the combustion resulted in crystalline as –prepared powders as well.

The diffraction X-ray results for powders prepared by both batches are presented in Figure 2. Full conversion to the single phase $LaCrO_3$ perovskite structure without need for subsequent calcinations was achieved for both powders. All the peaks of the XRD pattern were indexed and unit cell parameters were determined in various crystal systems using a standard computer program "PowdMult". Finally, a unit cell in orthorhombic system was selected for which the sum of differences in observed and calculated *d*-values i.e. $\sum \Delta d = \sum (d_{obs} - d_{cal})$ was found to be a minimum. The orthorhombic unit cell parameters, refined by least-squares methods are a=5.471 Å, b=5.507 Å and c=7.752 Å.

The scattered linear particle size (P_{hkl}) was calculated from strong and medium intensity peaks using Scherre's Equation.[9]

$$P_{hkl} = \frac{0.89\lambda}{\beta_{1/2}\cos(\theta_{hkl})} \tag{1}$$

where $\beta_{1/2}$ is half peak reflection. The average value was 44.14 nm and 38.03 nm for batch #1 and #2, respectively. This shows that powders prepared by combustion reaction with an excess of urea (fuel), batch #2, produces particles slightly smaller than that by on the urea stoichiometric, batch #1. This difference can be attributed to that the fact fuel amounts alter the energetics or exothermicity of the reaction and, consequently, the properties of final products. It was observed that for the rich-fuel mixture more flame were produced during combustion and also the combustion reaction took longer time before being totally completed.

The results of directed measurements of particle size distribution by sedimentation for powder produced by the batch #1 and batch #2 are $85.00 \pm 10\,nm$ and $74.00 \pm 15\,nm$, respectively. This narrow distribution of particle size of the both powders indicates that combustion reaction process yields fine uniform particles as a result of the rapid oxidation of nitrates at high temperature in situ in a viscous, homogeneous medium.

The relative density achieved for both powders was 96% which is higher when compared to 90% and 84% of those powders prepared by the Pechini method and mixing of oxides respectively.[10] Also the particle size of both powders prepared here when compared to those reported by Morelli,[10] shows that using a fuel-rich solution, smaller particle size powders are produced.

Although the diffraction X-ray patterns of the powders produced in the stoichiometric- and excess-urea, both as–prepared show no significant differences, the particle size is slightly different. Again it may be attributed to fact that fuel rich solution is highly exothermic, thus it alters the reaction, and consequently, the properties of final products.

The bulk density, as determined by Archimedes principle, is 95% and 97% of the theoretical value for pellets from powders batch #1 and batch #2, respectively. The densification of $LaCrO_3$ compacts at fairly low temperature (1600°C) can be attributed to the narrow particle-size distribution and the reactivity of fine-particle $LaCrO_3$ formed by the combustion method. This observation is remarkable because of the inherent problem involved in the sintering of $LaCrO_3$. It has been reported that above 1600°C in air chromium is volatilized and decomposing the stoichiometric crystalline phase.[4]

CONCLUSSIONS

Fine-particle of lanthanum chromite ($LaCrO_3$) was obtained by combustion reaction of the metal nitrate-urea mixtures. The exothermicity of the redox reaction in the metal nitrate-urea mixture produces powders with particle within a narrow particle-size distribution. For mixtures of excess fuel stoichiometric produced powders with particle size slightly smaller than those of urea (fuel) stoichiometric and more dense compacts as well. The powders had particle size below 1 μm and narrow size distribution making powders highly sinterable in air at 1600°C.

ACKNOWLEDMENTS

We would like to thanks the Universidade Católica de Goiás for the financial support. C.G.V. Neto is very greatfull to CNPq (Brazilian Financial Agency) for providing a scholarship under the PBIC program.

RFERENCES

[1] H. Kleinschmager and A. Reich, "Eletronic Conductors as Eletrodes of High Temperature Fuel Cells," *Z. Naturforsh*, **27** [2] 363-65 (1972).

[2] N.Q. Minh, "Ceramic Fuel Cell," *Journal of the American Ceramic Society*, **76** [3] 563-88 (1993).

[3] D.B. Meadowcroft, "Oxidation and Vaporization Process in Lanthanum Chromite," *American Ceramic Bulletin*, **58** [6] 610-15 (1979).

[4] H. Yokokawa, N. Sakai, T. Kawada and M. Dokiya, "Chemical Thermodynamic Considerations in Sintering of $LaCrO_3$-Based Perovskites," *Journal of the Electrochemical Society*, **138** 1018-27 (1991).

[5] L.W. Tai and P.A. Lessing, "Modified Resin-Intermediate Processing of Perovskite Powders," *Journal of Materials Research*, **7** [2] 502-09 (1992).

[6] S.S. Manoharan and K.C. Patil, "Combustion Synthesis of Metal Chromite Powders," *Journal of the American Ceramic Society*, **75** [4] 1012-15 (1992).

[7] S.R. Jain, K.C. Adiga, and V.R. Pai Vernecker, "A New Approach to Thermochemical Calculations of Condensed Fuel Oxidizer mixtures," *Combustion Flame*, **40,** 71 (1981).

[8] A.M. Wynne, "The Thermal Decomposition of Urea," *Journal of Chemical Education*, **64** [2] 180 (1987).

[9] B.E. Warren, "X-Ray Diffraction," Chapter 13, pp. 253, Dover Publications, INC., New York, 1990).

[10] M.R. Morelli, "Liquid Phase Sintering of Perovskites," *PhD Thesis, Univerisity of Oxford*, UK, pp. 85-130 (1995).

Fig.1. Combustion reaction steps: (a) liquid solution at the boiling stage.

Fig.1. Combustion reaction steps: (b) flame produced during gases ignition.

Fig.1. Combustion reaction steps: (c) fine product-crystallized $LaCrO_3$ powder.

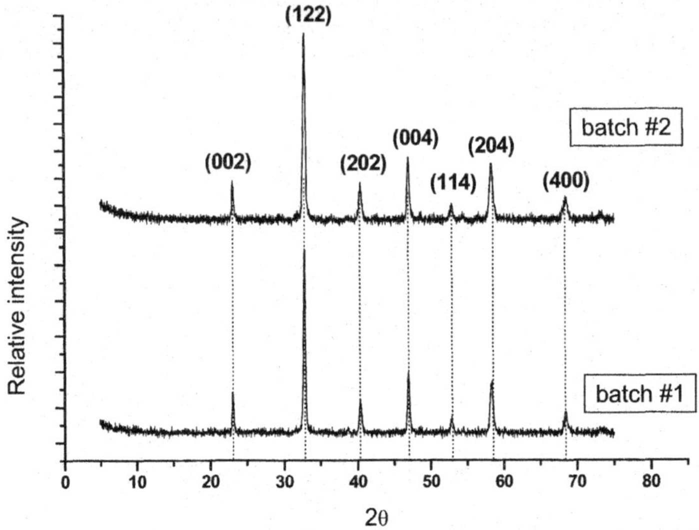

Fig.2 X-ray diffraction patterns of powders prepared by combustion reaction. Batch #1 fuel (urea) stoichiometric amount and batch # 2 an excess fuel (urea).

THE EFFECT OF PROCESSING CONDITIONS ON THE DIELECTRIC PROPERTIES OF LEAD MAGNESIUM NIOBATE-LEAD TITANATE CERAMICS

Barry Bender	Ming-Jen Pan
Code 6351	NOVA Research, Inc.
Naval Research Laboratory	1900 Elkin Street, Suite 230
Washington, DC 20375	Alexandria, VA 22308

ABSTRACT

A compositional series of lead magnesium niobate-lead titanate (PMN-PT) relaxor ferroelectrics have been produced via attrition milling and the columbite precursor method. Fine pure reactive PMN-PT powders were fabricated using careful processing. Sintering at 1200°C led to the formation of dense, phase pure, large-grained PMN-PT ceramics with high permittivity. Thermal annealing in flowing oxygen at 900°C optimized the dielectric properties of the PMN-PT ceramics.

INTRODUCTION

In its development of the all-electric ship the US Navy has made significant outlays in the technology of power electronics. However, passive components, especially filter capacitors, remain a limiting factor in power converter design. This is due to their low volumetric efficiency which causes them to be responsible for occupying 50 to 60% of the volume associated with today's state-of-the-art power converter. The ideal ceramic filter capacitor would consist of a high dielectric ceramic with good stability over a range of temperatures and voltages.

The high permittivity over a range of temperatures of various PMN-PT ceramics makes these lead-based perovskites possible candidates for filter capacitors.[1] Also due to their high diffuseness and reasonable sintering temperature PMN-PT ceramics have been the subject of many investigations for application in electronic ceramics. Research has shown that the synthesis of phase pure PMN-PT ferroelectrics is difficult and that their dielectric properties can be very sensitive to processing conditions. Compositional fluctuations within the starting powder can lead to the formation of the secondary pyrochlore phase which will directly lower the permittivity of the ceramic.[2] Calcination at temperatures too low can lead to incomplete reaction of the precursor powder. Sintering at too low of a temperature can prevent chemical homogenization of the PMN-PT ceramic.[3] Oxygen annealing can maximize dielectric properties through the reduction of a PbO-based grain boundary phase.[4] Therefore, to make a high dielectric relaxor ferroelectric ceramic, careful processing is needed.[5] This processing must lead to a phase pure, dense, large-grain ceramic that is chemically homogeneous. This research reports the effect of PbO precursor, calcination temperature, sintering temperature, and oxygen annealing on the dielectric properties of a series of PMN-PT ceramics and compares their properties to others reported in the literature.

EXPERIMENTAL PROCEDURE

A series of $(1\text{-}x)Pb(Mg_{1/3}Nb_{2/3})O_3\text{-}xPbTiO_3$ powders were prepared with the value of x =0.1 (90/10), x =0.2 (80/20), x =0.3 (70/30) and x =0.4 (60/40). The columbite precursor method[6] was used to inhibit the formation of the pyrochlore phase. The $MgNb_2O_6$ (MN) was made by ball-milling stoichiometric amounts of $MgCO_3$ and Nb_2O_5 (purity 99.9%, Alfa Aesar, Ward Hill, MA) in a solution of water and ammonia for 19 hours. The mixture was then dried and calcined at 1200°C for 4 hours. X-ray diffraction indicated that the resultant powder was not 100% MN so the ball-milling-calcine procedure was repeated three times until 100% MN was obtained. The final material was ball-milled and sieved through a 80 μm mesh to eliminate any large agglomerates.

Stoichiometric amounts of PbO or $PbCO_3$, TiO_2, and MN were blended into a purified water solution containing a dispersant (Tamol 901, Rohm and Haas, Philadelphia, PA) and a surfactant (Triton CF-10, Rohm and Haas, Philadelphia, PA). The resultant slurry was attrition-milled for one hour and then dried at 90°C. The powders were then calcined for 2 hours at various temperatures from 800 to 1000°C with 850°C being the standard calcination temperature. The calcined powders were then attrition-milled again, dried and sieved through a 45 μm mesh. XRD was done on all powders before compaction. A 2% PVA binder solution was mixed with the powder and the powder was re-sieved to eliminate any large agglomerates. The dried powder was uniaxially pressed into a disc 13 mm in diameter. The discs were placed on Pt foil surrounded by powder of similar composition and placed in a double-crucible setup. The discs were then sintered at 1150 or 1200°C (1200°C was the standard processing temperature) for 2h. Multiple discs were sintered at the same time and used for subsequent characterization and thermal annealing. Thermal annealing was carried out in flowing oxygen at 900 or 960°C for 6h (900°C was the standard temperature).

Material characterization was done on the disc and powders after each processing step. XRD was used to monitor the phase evolution for both the powders and the discs. Microstructural characterization was done on the fracture surfaces using scanning electron microscopy (SEM). To measure dielectric properties, sintered discs were ground and polished to achieve flat and parallel surfaces onto which palladium-gold electrodes were sputtered. The capacitance and dielectric loss of each disc were measured as a function of temperature (-60°C to 200°C) and frequency (100 Hz to 100 kHz) using an integrated, computer-controlled system working with a Hewlett Packard 4284A LCR meter.

RESULTS AND DISCUSSION

The Effect of PbO Precursor and Calcination Temperature

Figure 1 shows that 90/10 PMN fabricated with PbO had a maximum dielectric constant (Kmax) 20% greater than the 90/10 PMN fabricated from $PbCO_3$. XRD indicates that 90/10 and 80/20 PMN powder made using lead oxide was 100% PMN after calcination at 850°C, while the PMN powder made from lead carbonate showed the presence of pyrochlore and lead oxide (90/10- 2% pyrochlore and 80/20 PMN- 12% pyrochlore and 13% PbO). Calcination of the lead carbonate-based powders at higher temperatures to 1000° still led to the presence of free PbO (9%) and pyrochlore (18%). SEM of the 90/10 PMN lead carbonate-based powder showed the presence of large agglomerates ranging from 50 to 100 μm in size. The majority of the powder consisted of hard particles 2 to 3 μm in size. The 90/10 PMN lead oxide-based powder showed no signs of large agglomerates. Particles ranging from 2 to 3 μm in size were also observed, but they were sponge-like in appearance and consisted of finer particles typically ranging in size from 0.15 to 0.3 μm.

The above results indicate that lead oxide-based PMN powders were purer and more reactive which lead to an improvement of K_{max} by over 20% as compared to the dielectric properties of the lead carbonate-based PMN powders. According to Shrout et al.[7] reactive calcination should be occurring during 850°C calcination of the powders though it is at the high temperature end of the reactive calcination curve for PMN. However, in the case of the

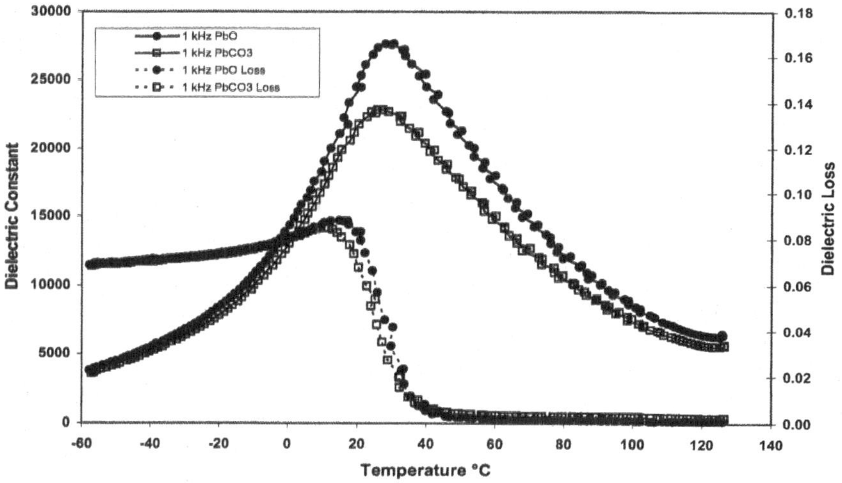

Fig. 1 Dielectric properties at 1 kHz of 90/10 PMN fabricated with lead oxide or lead carbonate.

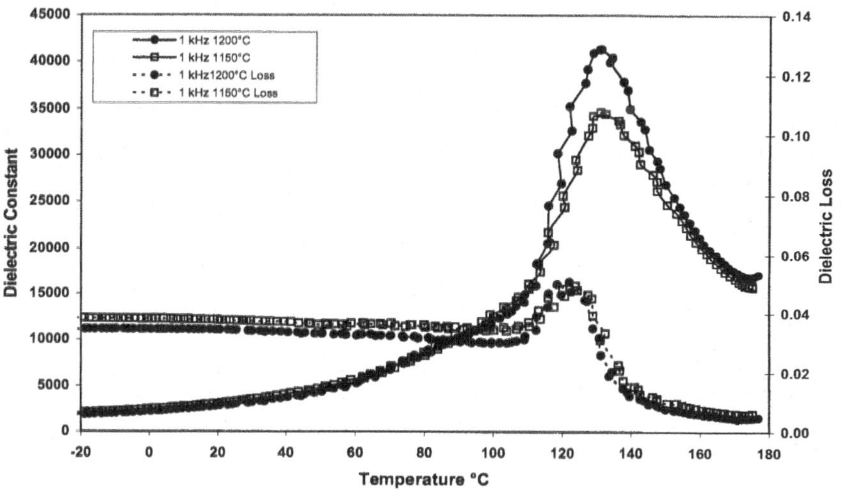

Fig. 2 Dielectric properties at 1 kHz of 70/30 PMN sintered at 1150°C or 1200°C.

carbonate-based powder it is entirely possible that 850°C may be too high for reactive calcination to take place leading to the formation of harder particles. The reason for this may be that lead carbonate may parallel magnesium carbonate in being more reactive.[8] As a result, the reactive calcination curve would be shifted to lower temperatures. The reactive calcination curve may be shifted to even lower temperatures because Kong et al.[9] have shown that for lead-based perovskites, that high energy milling can lead to compound formation at 100°C lower than conventionally-milled powders. This could lead to formation of particles that undergo little volumetric expansion during reaction and therefore would be harder to break up during subsequent attrition milling. As a result, the particle size of the lead carbonate-based powder would be larger and would lead to a lower density. This indeed was observed as the density of the lead oxide- based 90/10 PMN disc was 7.77 as compared to 7.66 g/cm^3 for the lead carbonate-based disc.

There are several possible reasons for the lower K_{max} of the lead carbonate-based ceramic. One is not grain size because the grain size for both cases is nearly identical (5.3 versus 5.4 μm). However, remnants of large agglomerates about 50 μm can be observed on the fracture surface of the carbonate-based PMN ceramics which were not observed for the lead oxide-based ceramic. This combined with the lower density would lower the permittivity of the ceramic. The other factor is the purity of the powder. Both Chen et al.[10] and Watanabe[2] show that lack of purity or compositional fluctuations in the starting powder can lead to significant lowering of the permittivity of PMN ceramics after sintering. In the case of the lead carbonate-based 90/10 PMN-PT ceramic after calcination at 850°C there was the presence of at least 2% pyrochlore unlike the lead oxide-based material which was 100% perovskite. Another indication that the powder was not as pure is that the carbonate-based material had a mass loss of 2.1% which was considerably higher than the lead oxide-based material which lost only 1.0% indicating that not all the lead oxide was reacted during calcination.

The Effect of Sintering Temperature (1150°C vs 1200°C)

Figure 2 shows the effect of sintering temperature on the dielectric properties of 70/30 PMN. Both the 1150 and 1200°C- sintered ceramics have similar T_{max} and similar losses. However, by increasing the sintering temperature by 50°C the K_{max} increased 20% from 34,600 to 41,300. Microstructural characterization indicated that the 1150°C-sintered ceramic was denser (7.84 vs 7.72 g/cm^3). However, the average grain size doubled from 3.3 to 6.6 μm upon sintering at 1200°C (see Fig. 3).

The substantial increase in K_{max} and grain size with a small increase in sintering temperature has been observed in other PMN systems. Ananta et al.[11] saw a dramatic improve in K_{max} as it increased 50% when the sintering temperature of pure PMN was increased from 1225 to 1275°C. Randall et al.[5] observed a 17% increase in K_{max} when the sintering temperature of 93/7 PMN was increased from 1150 to 1250°C. In regards to grain size, Liou et al.[12] measured an increase in grain size from 4.5 to 7.5 μm when the sintering temperature of PMN was raised only 50°C from 1220 to 1270°C.

Since the densities of the two samples are within 2% of each other, it appears that the doubling of the grain size is the major factor for the improvement of the permittivity of the 1200°C-sintered ceramic. A similar phenomenon was observed by Randall et al.[5] for a 93/7 PMN where K_{max} improved by 19% when the grain size increased from 3 to 6.5 μm, while the density remained the same for the two ceramics. The increase in dielectric constant with

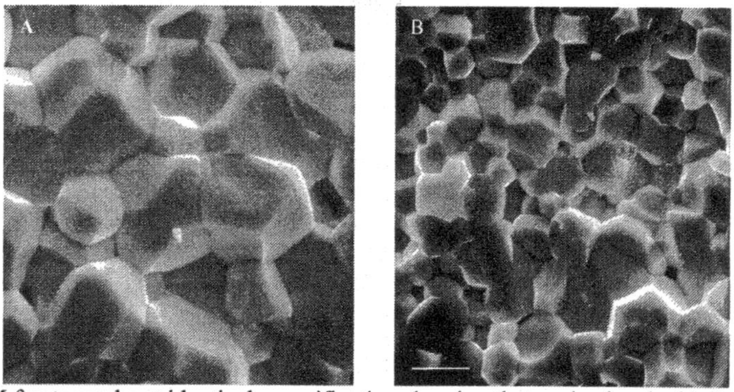

Fig. 3 SEM fractographs at identical magnification showing that grain size doubled on increasing the sintering temperature by 50°C (Fig. A- 1200°C and Fig. B- 1150°C, white marker is 3 μm).

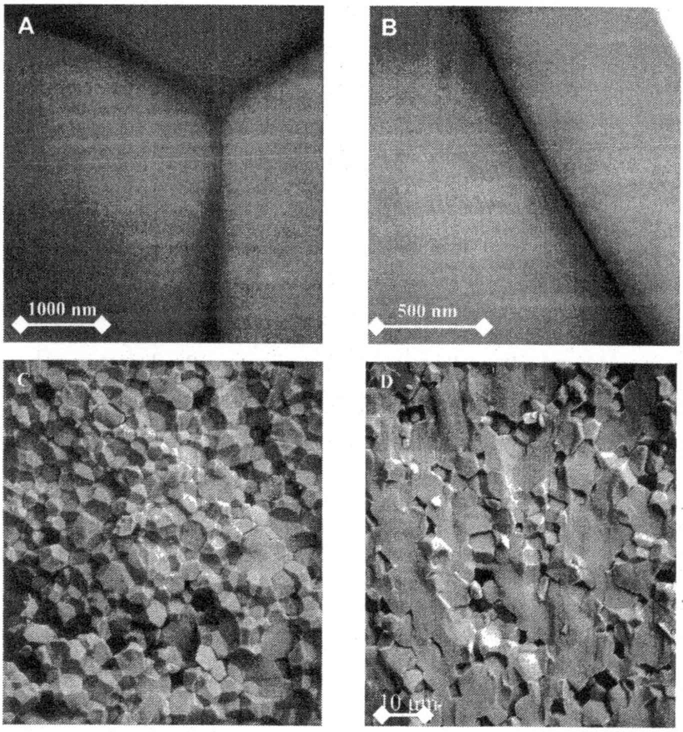

Fig. 4 SEM fractographs of the glassy grain boundary phase in PMN (A) as-sintered and (B) oxygen annealed at 900°C. Figs C and D are fractographs (same magnification) showing intergranular failure in as-sintered PMN (C) and a mix mode of fracture in annealed PMN (D).

increasing grain size has been attributed to various intrinsic and extrinsic effects. Both Wang et al.[13] and Randall et al.[5] have attributed the effect to the presence of a low dielectric PbO-rich glassy intergranular phase. Evidence of such of a phase was observed by SEM (see Fig. 4A) for each component of the PMN-PT series studied. The contribution of this parasitic phase to the overall permittivity of the PMN ceramic will decrease as the grain size increases which leads to the increase in dielectric constant with the increase in grain size.

Another possible contribution to the improvement in the permittivity of the 1200°C-sintered 70/30 PMN sample is chemical homogenization. Figure 2 shows that the dielectric response of the 1150°C-sintered sample is more diffuse as indicated by its slightly broader dielectric versus temperature response. This indicates that a greater degree of chemical homogenization is occurring at the higher sintering temperature.[3] Gupta et al.[14] saw a similar trend and attributed the improvement of the permittivity to the reduction of the low polarizable shell region around the PMN core which would lead to an increase in the dielectric constant of the overall material.

The Effect of Thermal Annealing

Figure 5 shows the effect of thermal annealing in flowing oxygen at 900°C for 6 h on the dielectric properties of 90/10, 80/20, and 70/30 PMN. For each composition K_{max} increased by 10 to 21 percent. With thermal annealing there was little weight loss ranging from 0.09 to 0.14 weight percent. Thermal annealing had no effect on density or measured grain size. XRD indicated that the samples were still 100% perovskite. However, there were changes in the type of fracture associated with thermal annealing. The unannealed samples broke primarily 90 to 95% intergranulary (see Fig. 4C) while the thermally annealed samples broke primarily in a mix mode of both transgranular and intergranular fracture (see Fig. 4D). Thermal annealing at 960°C (see Fig. 5) did not show substantial improvement in permittivity. The higher temperature annealing did lead to a greater weight loss of 2.1% and XRD detected the presence of pyrochlore.

The substantial increase in K_{max} with thermal annealing has been observed in other lead-based perovskite ferroelectric systems. Fan et al.[4] saw improvements by over 40% by thermal annealing in flowing oxygen. Randall et al.[15] use it to improve the dielectric and piezoelectric properties of PZT. The purpose of the thermal anneal is to eliminate the presence of the low dielectric PbO-rich glassy phase found between grain boundaries and triple point junctions in most lead-based perovskites.[5] However, unlike Ref. 5, the grain boundary phase distribution was not consistent from grain to grain in any of the as-sintered PMN-PT compositions. SEM observations indicated the presence of grain boundary phases that were not detectable by SEM to 200 nm in thickness (see Fig. 4A). After thermal annealing the glassy grain boundaries did appear to be reduced in thickness but were not totally eliminated (see Fig. 4B). This is similar to the results observed by Fan et al.[4] who observed via TEM a 50% reduction in the thickness of the grain boundary phase after thermal annealing. This correlates with the fractographs of the as-sintered and annealed PMN ceramics. The as-sintered samples fracture primarily in an intergranular mode (see Fig. 4C) due to the presence of the weak glassy phase present at the grain boundaries. Upon annealing, the thickness of the glassy phase is reduced making the grain boundary structure much stronger and as a result the annealed samples break in a mix mode of intergranular and transgranular fracture (see Fig. 4D).

Another possible contribution to the improvement in the permittivity of the thermally annealed PMN ceramics is homogenization. In our research as well as the research of Fan et al.[4]

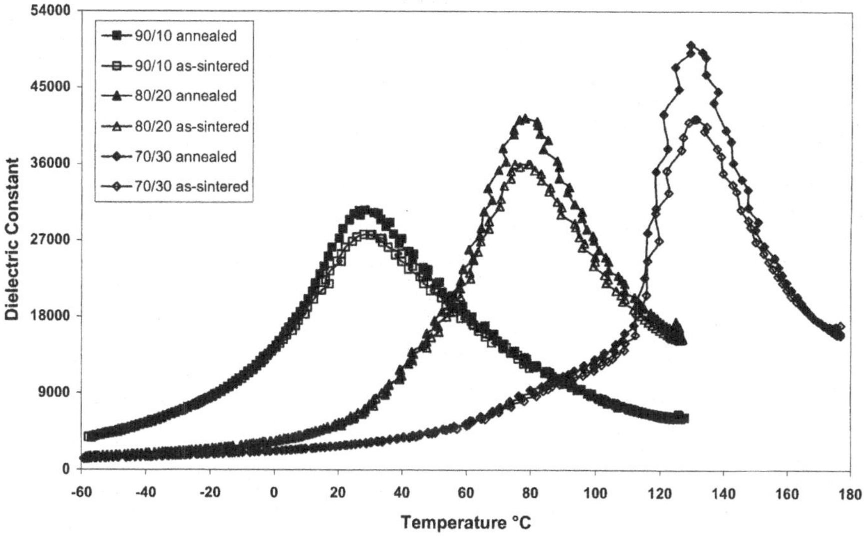

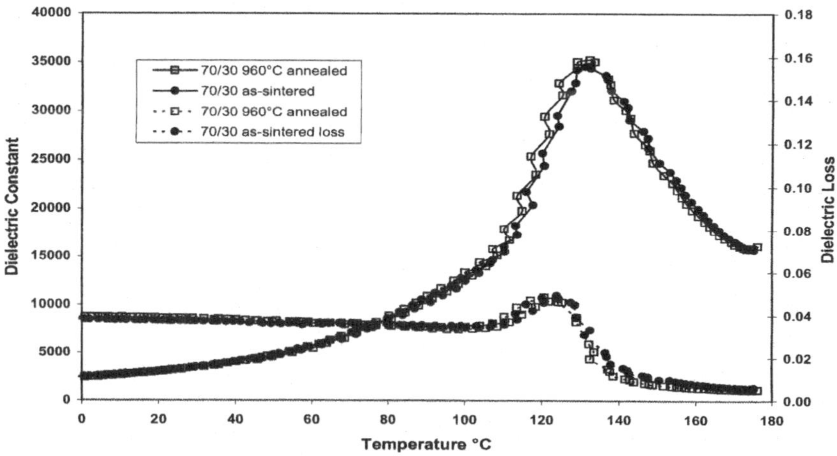

Fig. 5 Dielectric properties of PMN-PT ceramics annealed at 900°C (top) and 960°C (bottom).

thermal annealing leads to a less diffuse dielectric response with temperature. The broadness of the dielectric versus temperature response was reduced by 15% with thermal annealing which indicates a homogenization process is occurring. Papet et al.[16] proposed that with attrition milling that there may be a shell region on the grains where the crystallinity of the grain is less developed. This may be possible with high energy-milled powder because the process produces defects which make the powder more reactive.[17] Thermal annealing may heal these defects and reduce or eliminate the presence of this low polarizable phase shell region therefore increasing the overall dielectric constant of the ceramic.

Thermal annealing at 960°C did not lead to a substantial improvement in permittivity. XRD indicates the presence of pyrochlore which probably formed due to loss of PbO which is indicated by the jump in mass lost to 2.1% when annealing at the higher temperature. The presence of the low dielectric pyrochlore (*K*- 120) offsets most of the gains received by eliminating the glassy-PbO rich intergranular phase.

SUMMARY

Careful processing and annealing have produced PMN-PT ceramics with high permittivity. The K_{max} for the series of PMN-PT studied are 13 to 30 % higher than those reported in the literature.[3,8,18] The reason for the excellent dielectric properties is the fabrication of a fine pure reactive precursor powder via attrition milling and reactive calcination that was sintered at a high temperature to promote a dense, large-grained ceramic that showed no signs of the pyrochlore phase. Subsequent annealing optimized the permittivity of the PMN-PT ceramics.

Attrition milling, proper choice of precursors, and calcination temperature led to the development of a fine pure reactive PMN-PT powder. Since chemical fluctuations in the starting powder[2] can result in impure PMN-PT powder great care was taken to make sure that the columbite was 100% MN. Attrition milling was then use to mix the precursors well. Research indicated that lead oxide instead of lead carbonate led to purer PMN-PT powders. Research also showed that calcination at 850°C lead to a pure reactive powder. Attrition milling the powder after calcination lead to a powder with enhanced reactivity for several reasons. First of all, as observed by Shrout et al.[7] who measured a 10-fold reduction in particle size, attrition milling at this stage can lead to very fine powders. In our research particle sizes of 0.15 to 0.3 μm were observed after attrition milling. Also there were no signs of hard agglomerates which would reduce the sinterability of the powder. Finally, attrition milling is an higher energy milling process than ball milling and can possibly introduce structural defects and disorder into the powder which favors the diffusion process and therefore make the powder more reactive.[8,17] This higher reactive fine particle-sized powder should lead to enhanced densification.

However, pure PMN-PT starting powder doesn't guarantee high permittivity properties.[11] Research showed that by sintering at 1200°C instead of 1150°C that K_{max} increased by 20%. This increase was attributed to the doubling of the grain size and has been observed by other researchers.[5,14] There is also indications that the chemical homogeneity of the grains increase with increasing sintering temperature which would contribute to the improved dielectric constant of the PMN-PT ceramics.[3] Thermal annealing in flowing oxygen at 900°C optimized the final dielectric properties of the PMN-PT by 10 to 21% due to the reduction in thickness of the low-dielectric glassy phase found in between grains and at grain boundary triple points in the as-sintered PMN-PT ceramics.

REFERENCES

[1]D-H.Suh, D-H. Lee, and N-K. Kim, "Phase Developments and Dielectric/Ferroelectric Responses in the PMN-PT System," *J. Eur. Ceram. Soc.*, **22** 219-23 (2002).

[2]A. Watanabe, H. Haneda, Y. Moriyoshi, S. Shiraskaki, S. Kuramoto, and H. Yamamura, "Preparation of Lead Magnesium Niobate by a Coprecipitation Method," *J. Mat. Sci.*,**27** 1245-49 (1992).

[3]A.M. Scotch, A. Khan, T. Li, H.M. Chan, M.P. Harmer, "Chemical Heterogeneity in PMN-35PT Ceramics and Effects on Dielectric and Piezoelectric Properties," *J. Am. Ceram. Soc.,* **85** [12] 3018-24 (2002).

[4]H. Fan, G-T. Park, J-J. Choi, J. Ryu, and H-E. Kim, "Preparation and Improvement in the Electrical Properties of PZN-based Ceramics by Thermal Treatments," *J. Mater. Res.,* **17** [1] 180-85 (2002).

[5]C.A. Randall, A.D. Hilton, D.J. Barber, T.R. Shrout, "Extrinsic Contributions to the Grain Size Dependence of Relaxor Ferroelectric $Pb(Mg_{1/3}Nb_{2/3})O_3$:$PbTiO_3$ Ceramics," *J. Mater. Res.*, **8** [4] 880-84 (1993).

[6]S.L. Swartz and T.R. Shrout, "Fabrication of Perovskite Lead Magnesium Niobate," *Mater. Res. Bull.*, **17** [10] 1245-50 (1982).

[7]T.R. Shrout, P. Papet, S. Kim, and G. Lee, "Conventionally Prepared Submicrometer Lead-Based Perovskite Powders by Reactive Calcination," *J. Am. Ceram. Soc.,* **73** [7] 1862-67 1990.

[8]J-G. Baek, T. Isobe, and M. Senna, "Synthesis of Pyrochlore-Free $0.9Pb(Mg_{1/3}Nb_{2/3})O_3$-$0.1PbTiO_3$," *J. Am. Ceram. Soc.*, **80** [4] 973-81 (1997).

[9]L.B. Kong, J. Ma, W. Zhu, and O.K. Tan, "Reaction Sintering of Partially Reacted System for PZT Ceramics via a High-Energy Ball Milling," *Scripta Mater.*, **33** 345-50 (2001).

[10]J. Chen, A. Gorton, H.M. Chan, and M.P. Harmer, "Effect of Powder-Purity and Second Phases on the Dielectric Properties of Lead Magnesium Niobate Ceramics, " *J. Am. Ceram. Soc.*, **69** [12] C-188-90 1986.

[11]S. Ananta and N.W. Thomas, "Relationships between Sintering Conditions, Microstructure and Dielectric Properties of Lead Magnesium Niobate," *J. Eur. Ceram. Soc.,* 19 629-35 (1999).

[12]Y-C. Liou and K-H. Tseng, "Stoichiometric $Pb(Mg_{1/3}Nb_{2/3})O_3$ Perovskite Ceramics Produced by Reaction-Sintering Process, *Mat. Res. Bull.*, **38** 1351-57 (2003).

[13]H-C. Wang and W.A. Schulze, "The Role of Excess MgO or PbO in Determining the Microstructure and Properties of Lead Magnesium Niobate," *J. Am. Ceram. Soc.*, **73** [4] 825-32 (1990).

[14]S.M. Gupta, C.S. Harendranath, and A.R. Kulkarni, "Synthesis and Characterization of Lead Magnesium Niobate having Exceptionally High Dielectric Constant," *Ceram. Intl.*, **23** 191-96 (1997).

[15]C.A. Randall, N. Kim, J-P. Kucera, W. Cao, and T.R. Shrout, "Intrinsic and Extrinsic Effects in Fine-Grained Morphotropic-Phase-Boundary Lead Zirconate Titanate Ceramics," *J. Am. Ceram. Soc.,* **81** [3] 677-88 (1998).

[16]P. Papet, J.R. Dougherty, and T.R. Shrout, "Particle and Grain Size Effects on the Dielectric Behavior of the Relaxor Ferroelectric $Pb(Mg_{1/3}Nb_{2/3})O_3$, *J. Mater. Res.,* **5** [12] 2902-09 (1990).

[17]J. Xue, D. Wan, S-E. Lee, and J. Wang, "Mechanochemical Synthesis of Lead Zirconate Titanate from Mixed Oxides," *J. Am. Ceram. Soc.,* **82** [7] 1687-92 (1999).

[18]J. Kelly, M. Leonard, C. Tantigate, and A. Safari, "Effect of Composition on the Electromechanical Properties of $(1-x)Pb(Mg_{1/3}Nb_{2/3})O_3$-$xPbTiO_3$ Ceramics," *J. Am. Ceram. Soc.*, **80** [4] 957-64 (1997).

AN INVESTIGATION ON ELECTRIC FIELD-INDUCED MICROCRACKING AND TOUGHENING IN PIEZOELECTRIC CRYSTALS

Z. Xu and Y. Zhang
Department of Physics and Materials Science
City University of Hong Kong
Kowloon, Hong Kong, P.R. of China

ABSTRACT

Electric-field-induced microcracking and toughening in $Pb(Mg_{1/3}Nb_{2/3})O_3$-$PbTiO_3$ piezoelectric single crystals during electrically cycling were investigated using *in situ* transmission electron microscopy (TEM). The crack growth was found to depend on both cycling field strength and cycle number. The deflection of microcrack caused by defect was observed. The toughening induced by domain switching and formation of new domains around microcrack tip to resist the crack growth was directly demonstrated.

INTRODUCTION

Fracture behavior of ferroelectric materials has been extensively investigated in the past several decades. A number of fatigue mechanisms have been proposed from both microscopic and macroscopic points of view.[1-9] It is well known that for ferroelectric ceramics electrically induced cracking is different from the mechanically induced cracking in terms of their distinct fractography. A wedging action during the crack growth under an applied electric field is assumed based on the observation of intergranular fracture surface.[3,7] However, the dielectric breakdown mechanism was believed to contribute significantly to the electrically induced fracture considering the rough and melted fracture surface.[8] Recently, grain-boundary cavitation caused by dielectric breakdown and coalescence of microcracks were proposed to substantiate the micro-mechanism of electric-field-induced grain-boundary cracking behavior.[9] In all cases, the concentrated crack tip field is a key to explain the fracture behavior of ferroelectric materials.

It has been demonstrated from Vickers indentation-induced crack growth that the cracks parallel to poling direction are shorter than those perpendicular to poling direction. This fracture toughness anisotropy has widely been explained by reorientation of ferroelectric domains at the crack tip.[10-12] It is assumed that domain switching induced by the high stresses at the crack tip dissipates energy and thus gives rise to toughening behavior. However, no direct observation of toughening produced by the domain switching at the crack tip has been reported yet. In the present paper, the electric-field-induced crack growth characteristics and toughening produced by the domain switching at the crack tip in PMN-PT single crystals were examined by *in situ* TEM. The effects of the amplitude and cycle number of electric field as well as the impact of defects on the crack propagation were also investigated in order to gain a better understanding of fracture mechanism in ferroelectric single crystals.

EXPERIMENTAL PROCEDURE

The PMN-PT single crystal samples chosen for this study were grown by using a vertical Bridgman method with two nominal compositions of 0.70PMN-0.30PT and 0.65PMN-0.35PT. The TEM specimens were prepared from the slice of (001) single crystals by ultrasonically cutting, polishing, dimpling. After that, two semicircular electrodes with 150μm gap wide were sputtered on the undimpled side of the 3mm disc. The central part of the disc was thinned to electron transparency with Ar ion beam from the unelectroded side. The specially designed double tilt sample holder was used to apply bipolar electric field to the TEM specimens. A Trek Model 609E-6-FG high-voltage function generator was used to supply the cycling electric field with different amplitudes and frequencies. *In situ* TEM observations were carried out on a Phillips CM-20 microscope. Figure 1 shows a schematic diagram of TEM specimen configuration for the in situ TEM study on field-induced microcracking. Since the applied field was perpendicular to the optical axis of the TEM, it would cause electron beam deflection. The beam deflection could be easily compensated for by adjusting beam deflection on the microscope. On the other hand, the direction of applied electric field with respect to the interesting features in the specimen, such as microcracks and domain boundaries, can be easily determined by checking the deflection of a focused electron beam in the TEM specimen. Normally, TEM images could be mornitored and recorded when a static electric field was applied to the specimen. For a cyclic field TEM images could be recorded only after the field was removed from the specimen.

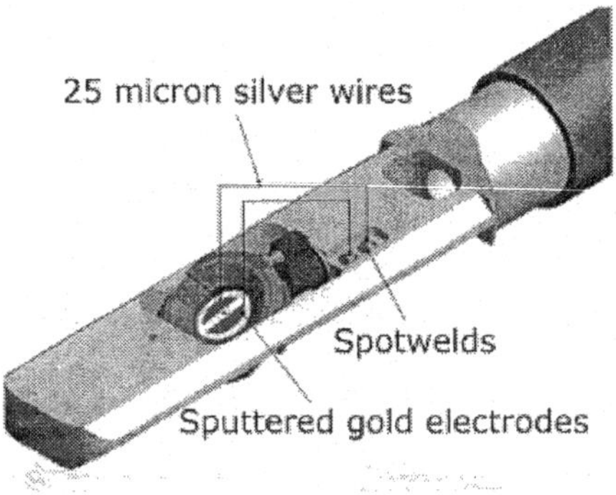

Fig. 1 Schematic diagram of specially designed double tilt TEM specimen holder.

RESULTS AND DISCUSSION

The TEM specimen was electrically cycled starting at a small field strength and then increasing the field strength gradually. After the specimen was subjected to a certain field level for a period of time, the cycling field was removed in order to see whether any microcracks were initiated in the specimen. No microcracks were detected until the field strength reached a critical value. For the 0.70PMN-0.30PT crystal specimen a microcrack was found to be emanated from the perforation at the center of the TEM specimen after the applied field strength reached to 30kV/cm (f=1Hz). Figure 2 are TEM micrographs showing growth of this newly field-induced microcrack under increased electric field strength (E = 30 to 36 kV/cm). After the specimen had been electrically cycled for 300 times at 30kV/cm, the crack was extended to about 2μm in length and the crack tip was marked in Figure 2(a). The microcrack is perpendicular to the direction of applied field and thus it propagates, as expected. It should be mentioned that the microcrack propagates across wavy domain boundaries. With further increasing the number of switching cycle at E=30kV/cm, no further crack extension was observed. However, as the electric field strength was increased to E=33kV/cm and cycled for another 300 times, a cloud-shaped segment ahead of the crack tip was developed, which was accompanied by a further opening of the crack tip, as shown in Figure 2(b). This indicates that the microcrack growth was arrested at some defect like cloud-shaped segment here at E=33kV/cm. No further microcrack growth was observed with increasing the number of cycle at this field level. However, after increasing the amplitude of the cyclic electric field, a triangle-shaped material or defect was found to be removed at the crack tip during crack extension, as shown in Figure 2(c), and then the crack tip deflected slightly and propagated again after the specimen was subjected to another electrically cycling. The crack in a perfectly homogenous sample under applied stress usually propagate straight. But the introduce of defects will generate electric field singularity and lead to the disturbance of the crack growth path.[13] From the above experimental observation, the triangle-shaped defect might be impurity inclusions that were introduced during crystal growth. In this case the crack deflection was mainly caused by defects in the single crystal rather than by the external driving force. The strong strain compatibility around the defect may drive the crack to propagate along a plane that satisfying the maximum strain criterion.[14]

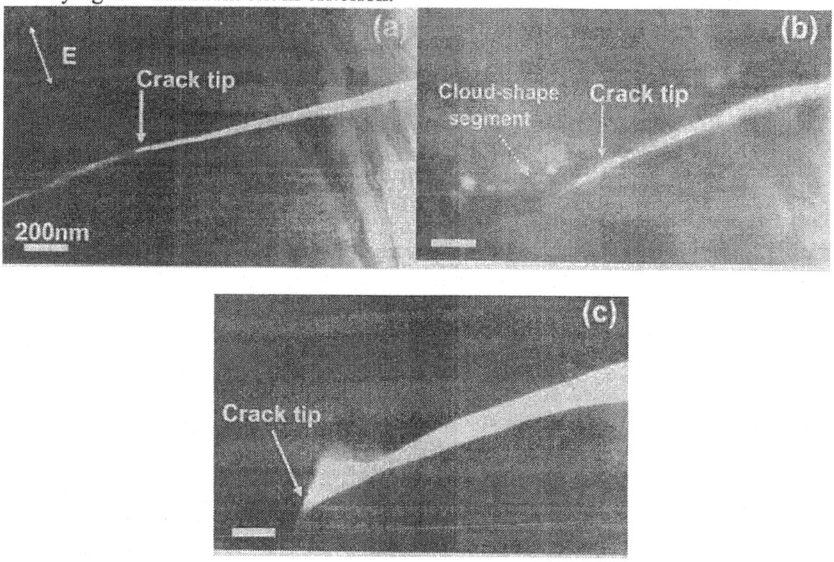

Fig. 2 TEM micrographs showing the microcrack propagation with increasing cyclic electric field strength (f = 1Hz) for $0.70Pb(Mg_{1/3}Nb_{2/3})O_3$-$0.30PbTiO_3$ crystal specimen. a) after 300 cycles at 30kV/cm; b) after 300 cycles at 33kV/cm; c) after 900 cycles at 36kV/cm.

The further propagation of the deflected crack at an field of 36kV/cm with increasing cycle number is illustrated in Figure 3. It is noted that cavities were nucleated near the crack tip first (Figure 3(a)) and then were joined together, extending the previous microcrack under continued electric cycling (Figure 3(b)). This nucleation of cavities led before microcrack propagation was previously reported in *in situ* TEM study of piezoelectric lead zirconate titanate ceramics.[9] The similar phenomena was for the first time observed in piezoelectric PMNPT single crystal in this work. On the other hand, it can be seen that with increasing the field cyclic number, the crack undergone a transitorily stable growth stage (Figure 3(a)) and advanced at a growth rate of about 10^{-9} m/cycle (Figure 3(b)). Further cycling at this field strength led to a sudden catastrophic fracture of the specimen. This kind of discontinuous fatigue crack growth was frequently observed in single crystal samples even under optical microscopy observation.[15]

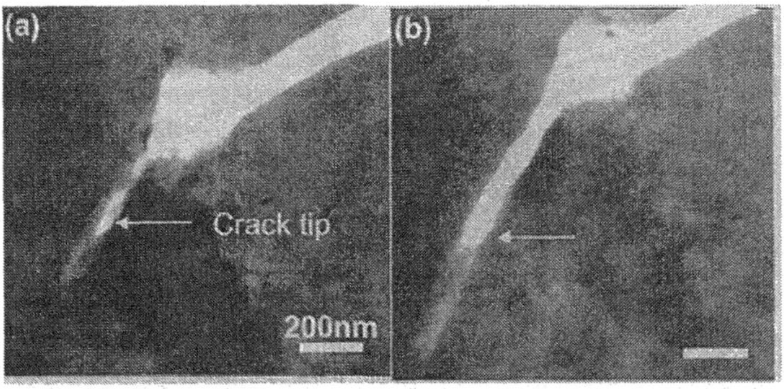

Fig. 3 TEM micrographs showing the microcrack growth with increasing switching cycles at an electric field (36kV/cm, f=1Hz) for $0.70Pb(Mg_{1/3}Nb_{2/3})O_3$-$0.30PbTiO_3$ crystal specimen. a) after 1200 cycles and b) after 1260 cycles.

To explain the electric-field-induced crack growth behavior, let us consider the complete constitutive framework where the stress intensity factor for a microcrack of length 2a is given by[16]

$$K_I \approx 0.25 \frac{K_E Y \gamma_s}{E_c} \tag{1}$$

Where Y is the Young's modulus, γ_s is the saturated strain, and E_c is the coercive field. K_E is the electric intensity factor and expressed as

$$K_E = E\sqrt{\pi a} \tag{2}$$

Where E is the applied electric field strength. Taking the typical data of Y=10GPa, γ_s=0.01, a=2μm, E=30 kV/cm, E_c=3kV/cm for the 0.70PMN-0.30PT single crystal, then the stress intensity factor for the crack reaches 0.63 MPa $m^{1/2}$. This value falls in the range of the fracture toughness of the single crystal.[17] Substituting Eq. (2) into Eq. (1) indicates that the stress intensity factor increases with the applied electric field strength. The experimentally observed crack propagation as a function of the applied electric field amplitude can be accounted for by this complete constitutive framework approach. Nevertheless, such formulas are not sufficient to fully describe the cycle number dependence of the crack growth. Therefore, the above discussion is still qualitative. In addition, the exact value of the local applied electric field at the crack tip in our *in situ* TEM experiment is hard to determine. The presence of discontinuous propagation of cracks suggests that there are other contributions to the electric-filed-induced crack process. The similarity between our observations in Figure 3 and those published previously[9] indicates that an underlying field-induced cavitation even in single crystal may well be responsible for the fracture of the ferroelectric materials. The formation of the cloud-shape zone as well as the removal of defect also demonstrate that the generation of high energy and high stress concentrations by cyclic excitation accumulate at the crack tip. There is hence an implication that one kind of stress corrosion cracking mechanism may also be expected to play a relevant role in the fracture process.[18]

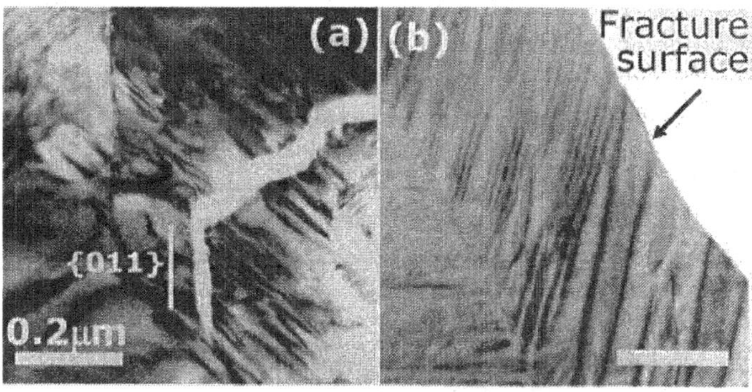

Fig. 4 TEM micrographs showing cross-domain boundary microcrack propagation (a) and freshly fractured surface of $0.70Pb(Mg_{1/3}Nb_{2/3})O_3$-$0.30PbTiO_3$ crystal specimen.

It should be mentioned that previous *in situ* TEM study on large-domain sized PMNPT single crystals revealed that field-induced microcracks in piezoelectric crystals prefer domain boundary as the crack path, which indicates that the effect of incompatible strain at the domain boundary outweighed that of field intensification on determining the microcrack propagation path.[19] However, in this work the studied PMNPT crystal has a smaller domain size (domain width < 0.1μm) than that in the above mentioned work (domain width ~ 0.3μm), microcracks were often observed to propagate across domain boundary, instead of along the boundary, as shown in Figure 2(a). Another example is illustrated in Figure 4, which are two micrographes obtained from another 0.70PMN-0.30PT crystal specimen before and after occurrence of a localized fracture. Figure 4(a) clearly reveals that the microcrack propagates across the nanometer-sized domain boundaries and Figure 4(b) shows the frashly-fractured surface of the TEM specimen. In this case, the effect of field intensification at the crack tip outweighed that of incompatible strain at the domain boundary on determining the microcrack propagation path. A more detailed study of electric field-induced microcrack propagation across domain boundary in piezoelectric PMNPT crystals will be published elsewhere.

During *in situ* TEM studies of electric field-induced microcrack propagation, an interesting toughening phenomenon was directly observed for the first time, as illustrated in Figure 5, which is a TEM micrograph of a 0.65PMN-0.35PT single crystal specimen recorded just before a sudden catastrophic fracture of TEM specimen. The 0.65PMN-0.35PT crystal specimen has a large stripe-like domain structure (submicron in width) as shown and electric field-induced microcrack was found to propagate along {011} domain boundaries, as expected. The specimen had been electrically driven at a field of 6.5kV/cm (f=0.3Hz) for more than several hundreds of cycles. The microcrack had propagated about 2 μm along the domain boundary until accumulated crack tip stress reseached at a critical level, the crack suddenly deflected original path and propagated across domain boundary, as shown (marked by a small arrow). After recording this image, the specimen was driven by the same cyclic field for a few more hundred of cycles and suddenly fractured during field cycling. Before catastrophic failure, a large strain field was developed around the microcrack tip, as indicated by the presence of strong strain contrast. It is believed that the highly localized and intensified stress field around the crack tip caused the deflection of microcrack propagation direction. This intensified field also caused local domain switching, as indicated by widening of domains that are close to the deflected microcrack (these domains usually have a near constant width) as well as formation of small new domains around the deflected microcrack tip as marked by a large arrow. The domain switching and formation of new domains around the crack tip absorb energy that leads to the arrest of further microcrack growth or toughening. This toughening behavior was directly observed at a microscopic level for the first time. The accommodation process envolved in toughening could be achieved by a deformation twinning mechanism.[12] A more systematic study of fracture and toughnening mechanism in piezoelectric single crystals is in progress.

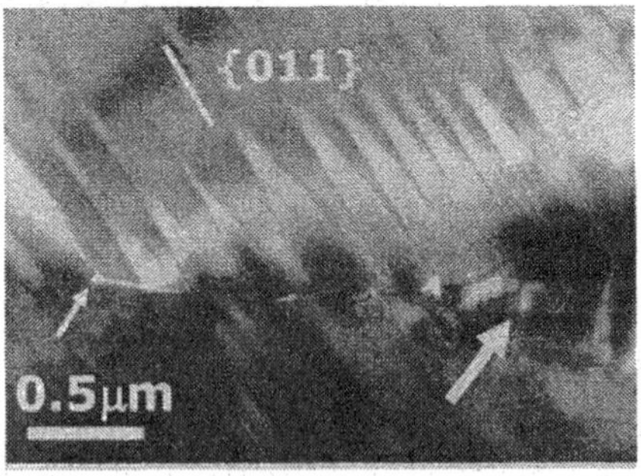

Fig. 5 TEM micrograph of a 0.65PMN-0.35PT single crystal specimen recorded just before occurrence of a sudden catastrophic fracture of TEM specimen.

CONCLUSIONS

Microcracking and toughening behavior in PMN-PT piezoelectric single crystals under electrical loading have been investigated using the unique *in situ* TEM technique. The results have demonstrated that the electric field-induced crack growth exhibits a strong electric field strength dependence. The crack growth with increasing switching cycles at a fixed field strength and the defect-induced crack deflection were directly observed. Domain switching and formation of new domains near microcrack tip were demonstrated for the first time, which resulted in local toughening to resist the crack growth.

ACKNOWLEDGMENT

This work was supported by a grant from the Research Grant Council of the Hong Kong Special Administrative Region, China [Project No. 9040643]. One of the authors Dr. Zhang is grateful to Dr. William S. Oates from the Georgia Institute of Technology for helpful discussions.

REFERENCES

[1]J. Nuffer, D.C. Lupascu and J. Roedel, "Damage Evolution in Ferroelectric PZT Induced by Bipolar Electric Cycling," *Acta Materialia*, **48**, 3783-3794 (2000).

[2]T. Zhu and W. Yang, "Fatigue Crack Growth in Ferroelectrics Driven by Cyclic Electric Loading," *Journal of the Mechanics and Physics of Solids*, **47**, 81-97 (1999).

[3] H. Cao and A.G. Evans, "Electric-Field-Induced Fatigue Crack Growth in Lead Zirconate Titanate," *Journal of American Ceramic Society*, **77**, [7] 1783-1786 (1994).

[4]C. S. Lynch, W. Yang, L. Collier, Z. Suo and R.M. McMeeking, "Electric Field Induced Cracking in Ferroelectric Ceramics," *Ferroelectrics*, **166**, 11-30 (1995).

[5] S. J. Kim, and Q. Jiang, "Microcracking and Electric Fatigue of Polycrystalline Ferroelectric Ceramics," *Smart Matererials and Structures*, **5**, [6] 321-326 (1996)..

[6]W. Yang, *Mechatronic reliability*, Springer, Berlin, Heidelberg, 2002.

[7]C. S. Lynch, L. Chen, Z. Suo, R.M. McMeeking and W. Yang, "Crack Growth in Ferroelectric Ceramics Driven by Cyclic Polarization Switching," *Journal of Intelligent Material Systems and Structures.* **6**, [3] 191-198 (1995).

[8]T. H. Wang and T. Y. Zhang, "Electrical Fracture Toughness for Electrically Conductive Deep Notches Driven by Electric Fields in Depoled Lead Zirconate Titanate Ceramics," *Applied Physics Letters*, **79**, [25] 4198-4200 (2001).

[9]X. L. Tan and J.K. Shang, "In-situ Transmission Electron Microscopy Study of Electric Field-induced Grain Boundary Cracking in Lead Zirconate Titanate," *Philosophical Magazine* **A82**, [8] 1463-1478 (2002).

[10]K. Mehta and A.V. Virkar, "Fracture Mechanisms in Ferroelectric-Ferroelastic Lead Zirconate Titanate (Zr:Ti = 0.54:0.46) Ceramics," *Journal of American Ceramic Society*, **73**, [3] 567-574 (1990).

[11]A. Kolleck, G.A. Schneider and F.A. Meschke, "R-curve Behavior of $BaTiO_3$- and PZT Ceramics under the Influence of An Electric Field Applied Parallel to the Crack Front," *Acta Materialia*, **48**, [16] 4099-4113 (2000).

[12]M. J. Reece and F. Guiu, "Toughening Produced by Crack-tip-stress-induced Domain Reorientation in Ferroelectric and/or Ferroelastic Materials," *Philosophical Magazine* **A 82**, [1] 29-38 (2002).

[13] S. L. dos Santos e Lucato, H.A. Bahr, V.B. Pham, D.C. Lupascu, H. Balke, J. Roedel, and U. Bahr, "Electrically Driven Cracks in Piezoelectric Ceramics: Experiments and Fracture Mechanics Analysis," *Journal of the Mechanics and Physics of Solids*, **50**, [11] 2333-2353 (2002).

[14]J.K. Shang and X.L. Tan, "A Maximum Strain Criterion for Electric-field-induced Fatigue Crack Propagation in Ferroelectric Ceramics," *Materials Science and Engineering* **A301**, 131-139 (2001).

[15]Liu, T.Q., Oates, W.S., Wan, S., and Lynch, C.S., 2003, unpublished.

[16]T. H. Hao, X. Gong and Z. Suo, "Fracture Mechanics for the Design of Ceramic Multilayer Actuators," 1996, *Journal of Mechanics Physics Solids*, **44**, 23-48 (1996).

[17]Y. Hosono, K. Harada, S. Shimanuki, S. Saitoh and Y. Yamashita, "Crystal Growth and Mechanical Properties of $Pb[(Zn_{1/3}Nb_{2/3})_{0.91}Ti_{0.09}]O_3$ Single Crystal Produced by Solution Bridgman Method," *Japan Journal of Applied Physics*, **38**, 5512-5515 (1999).

[18]Y. Wang, W.Y. Chu, K.W. Gao, Y.J. Su and L. J. Qiao, "Delayed Fracture of Lead Zirconate Titanate Ferroelectric Ceramics under Sustained Electric Field," *Applied Physics Letters*, **82**, [10] 1583-1585 (2003).

[19]X. Tan, Z. Xu and J.K. Shang, "Direct Observations of Electric Field-induced Domain Boundary Cracking in <001> Oriented Piezoelectric $Pb(Mg_{1/3}Nb_{2/3})O_3$-$PbTiO_3$ Single Crystal," Appl. Phys. Lett., **77** [10] 1529-1531 (2000).

Crystal Chemical and Physical Related Issues, Synthesis and Crystal Growth

HOMOGENEITY REGION AND CRYSTAL STRUCTURE OF Sr AND Mg CONTAINING $LaGaO_3$ AT TEMPERATURES BETWEEN 1100°C AND 1500°C IN AIR

Peter Majewski,
Ian Wark Research Institute
University of South Australia
Mawson Lakes Blvd., Mawson Lakes
SA 5095, Australia

Michael Rozumek
Webasto AG
Speicherstrasse 3-4
17033 Neubrandenburg
Germany

Fritz Aldinger
Max-Planck-Institut für Metallforschung
70569 Stuttgart, Germany

ABSTRACT

Phase stability and crystal strucutre studies were performed within the quasi-ternary system $LaGaO_3$-$SrGaO_{2.5}$-"$LaMgO_{2.5}$". Emphasis was cast on the temperature dependence of the homogeneity region of $La_{1-x}Sr_xGa_{1-y}Mg_yO_{3-\delta}$ perovskite solid solutions. Isothermal sections were determined at 1100°, 1250°, 1400°, and 1500°C in a static air atmosphere. The single-phase homogeneity region was found to considerably diminish with decreasing temperature, indicating a dependence of the solid solubility of Sr and Mg on the temperature. The cubic perovskite phase was found to exist only at elevated temperatures and for high Sr and Mg amounts. Sample preparation was performed by the mixed-oxide process as well as by a modified combustion synthesis.

INTRODUCTION

In the wake of research findings on the superior ionic conductivity in Sr- and Mg- doped $LaGaO_3$ compounds (i.e., solid solutions of $La_{1-x}Sr_xGa_{1-y}Mg_yO_{3-\delta}$, LSGM) published by Goodenough and co-workers[1-7], Ishihara *et al.*[8-13] in 1994, and Huang *et al.*[14] in 1996, extensive research activities have been launched to evaluate the usefulness of this novel class of perovskite materials as electrolytes for reduced-temperature solid oxide fuel cells (RTSOFC)[15-21].

While functional properties of LSGM, such as electrical conductivity, electrode kinetics, overall cell performance, etc. have been thoroughly studied by various authors[2,9,11,17,22], there has nonetheless been a long-standing need for a detailed phase diagram study in the system La_2O_3-SrO-Ga_2O_3-MgO, and investigations of the crystal structure of LSGM.

Several authors [4-6,23] thoroughly investigated the conditions under which single phase LSGM can be obtained. However, when the LSGM compound is considered to be used as a solid electrolyte over extended periods of time at a temperature reasonably different from the fabrication temperature (*e.g.*, in RTSOFCs), it is a crucial requirement to gain a comprehensive knowledge of the temperature dependence of the homogeneity region of LSGM compositions. A good understanding of the underlying phase diagram is a prerequisite, when ageing phenomena such as decomposition reactions are to be assessed for LSGM electrolytes in long term use at reduced temperatures.

Isothermal sections through the quasi-ternary systems La_2O_3-SrO-Ga_2O_3, La_2O_3-Ga_2O_3-MgO, SrO-Ga_2O_3-MgO for 1400°C in air were just being published by the present authors elsewhere[24,25]. This paper instead focuses on the experimental assessment of the quasi-ternary sytem $LaGaO_3$-$SrGaO_{2.5}$-"$LaMgO_{2.5}$", placing emphasis on the determination of the homogeneity region of the perovskite solid solution as a function of temperature.

$LaGaO_3$ has a Perovskite-type structure (ABO3) with La entering the A-site and Ga entering the B-site within the oxygen octrahedron. Pure $LaGaO_3$ is orthorhombic, space group Pbnm, with a = 5.5245 Å, b = 5.4922 Å, and c = 7.7740 Å [26]. At about 200 °C a phase transformation from orthorhombic to rhombohedral, space group $R\bar{3}c$, had been observed [26,27]. The crystal structure of Sr and Mg rich $LaGaO_3$ has recently been reported to be cubic, space group $Pm\bar{3}m$ and a = 3.9103 Å [4]. However, recent investigations of the structure of pure as well as Sr and Mg substituted $LaGaO_3$ ($La_{0.9}Sr_{0.1}Ga_{0.2}Mg_{0.2}O_3$) using high temperature neutron scattering reveal orthorhombic structure, space group Pnma with a = 5.4908 Å, b = 7.7725 Å, and c = 5.5227 Å, for pure $LaGaO_3$, and space group Imma with a = 5.5179 Å, b = 7.8200 Å, and c = 5.5394 Å for Sr and Mg substituted $LaGaO_3$ at room temperature. At temperatures of 800 and 1000 °C a rhombohedral structure, space group $R\bar{3}c$, for the pure compound and a cubic modification, space group $Pm\bar{3}m$, for the substituted phase was reported [27].

These observations indicate phase transformations of $LaGaO_3$ to take place with increasing Sr and Mg content as well as temperature. In this work the crystal structure of $LaGaO_3$ has been determined using X-ray powder diffraction and Rietveld analysis with respect to increasing Sr and Mg contents.

EXPERIMENTAL

(1) Sample Preparation

Powder and bulk samples having overall compositions of $La_{1-x}Sr_xGa_{1-y}Mg_yO_{3-\delta}$ with x=0-0.3 and y=0-0.25 were primarily prepared by a modified combustion technique via aqueous solutions. Samples for phase equilibria studies performed at temperatures of 1100°, 1250°, 1400°, and 1500°C were obtained by this route. For sample preparation at high temperatures, additionally conventional solid-state reactive firing was employed, i.e. the conventional mixed-oxide route. Samples obtained by both methods were identical with respect to their chemical and phase composition after annealing at 1400° and 1500°C, verifying the high-temperature compatibility of the two processing routes with one another.

Combustion Synthesis: LSGM powders were prepared by a modified glycine-nitrate-process[28], using urea as a fuel. Previous to this study, related processing techniques had been utilized for the preparation of this material by Stevenson *et al.*[29-31] and Mathews *et al.*[32] In our work, $La(NO_3)_3{\cdot}6H_2O$ (99.0+%, Sigma-Aldrich, Steinheim, Germany), Sr-Acetate (99.5+%, Sigma-Aldrich, Steinheim, Germany), $Ga(NO_3)_3{\cdot}xH_2O$ (99.999%, Sigma-Aldrich, Steinheim, Germany), $Mg(NO_3)_2{\cdot}6H_2O$ (99.0+%, Sigma-Aldrich, Steinheim, Germany), and urea (Merck, Darmstadt, Germany) were used as raw materials. For each sample, appropriate volumes were taken from defined aqueous stock solutions (prepared with bi-distilled and de-ionized water) of the corresponding metal salts and thoroughly mixed. The amount of urea added was calculated following the method derived by Jain *et al.*[33] for complete combustion. In this route of preparation, urea not only serves as an oxidizer during combustion but also as a complexant, avoiding the undesired precipitation of single cation-salts upon gelling. This effect helps to keep the cations homogenously distributed in the full course of powder preparation. For the present synthesis, solutions rich in fuel (combustion parameter $\Phi_e \approx 2$) were prepared. Excess water was evaporated from each solution until gelling, before the mixtures were transferred into covered alumina crucibles and placed into a box furnace with a static air atmosphere. Thermal treatment was carried out in a two-step process by first bringing samples to the ignition temperature, followed by cold-isostatic compaction at 625 MPa for 60 s, and subsequently to the annealing temperature chosen for a particular phase study. Heating rates of 5 K/min were chosen. Annealing was performed varying over the range of 50-100 h and in a static air atmosphere. All samples were furnace-cooled to room temperature with cooling rate of 2-10 K/min.

Solid State Synthesis: Starting powders of La_2O_3 (99.99%, Sigma-Aldrich, Steinheim, Germany), $SrCO_3$ (98+%, Sigma-Aldrich, Steinheim, Germany), Ga_2O_3 (99.99+%, Sigma-Aldrich, Steinheim, Germany), and MgO (98+%, Merck, Darmstadt, Germany) were first dried at 120°C. Powders were then weighed and mixed in appropriate amounts to give the desired overall compositions. Powder mixtures were mixed in iso-propanol (Merck, Darmstadt, Germany) in an agate orbital ball mill and sieved down to 100 μm, before cold-isostatic pressing at 625 MPa for 60 s into small pellets (diameter ≈ 5-10 mm, height ≈ 5-10 mm). Pellets were calcined at 1200°C for 8 h in air. After a subsequent grinding step in a WC ball mill (average particle size < 160 μm) and repeated cold-isostatic pressing under afore-said conditions, bulk samples were finally annealed at 1400° and 1500°C for 12 h in a static air atmosphere. Heating, and cooling conditions were the same as given above.

(2) Sample Characterization

Qualitative phase distributions in the annealed samples were studied by powder X-ray diffraction analysis using $CuK_{\alpha 1+2}$ radiation with 40 kV acceleration voltage, 25 mA filament current (D5000, Kristalloflex, Siemens, Karlsruhe, Germany). θ-2θ scans with a position sensitive proportional counter were recorded at a step size of 2θ=0.008° and a step time of 1.5 s. Sample preparation was performed by crushing and grinding annealed pellets in an agate mortar, obtaining a fine-grained powder with a particle size < 36 μm, which for XRD analysis was subsequently layed down onto the substrate as a dense and smooth layer. As-ground samples were not subjected to a further annealing step for relaxation of residual grinding stresses, as not to aid decomposition reactions of the non-equilibrium samples.

SEM was used for microstructural characterization, utilizing a W anode at typically 20 kV acceleration voltage (REM S200, Cambridge Instruments, UK). Chemical analysis was performed on an attached EDX spectrometer (LINK AN 10 000, Pentafet detector). For specimen preparation, as-received bulk samples were first embedded in a carbon containing epoxy polymer and subsequently ground and polished in a multi-step procedure, finishing with a ¼ μm diamond suspension on a soft cloth (DP-NAP, Struers, Copenhagen, Denmark).

For verifying the desired overall composition of each sample, as-received powders and bulk samples were subjected to quantitative chemical analysis, using ICP-OES (Spectroflame, Germany).

Crystal structure characterization was performed by room temperature powder X-ray diffraction (XRD) analysis using $CuK_{\alpha 1+2}$ radiation with 40 kV acceleration voltage, 25 mA filament current (D5000, Kristalloflex, Siemens, Germany). θ-2θ scans with a position sensitive proportional counter were taken at a step size of $2\theta = 0.008$ ° and a step time of 1.5 s. Sample preparation was carried out by crushing and grinding annealed pellets in an agate mortar, obtaining a fine grained powder with a small particle size (< 36 μm), which for XRD analysis was subsequently layed down onto the substrate as a dense and smooth layer. Diffraction data manipulation entailed smoothing, background and $CuK_{\alpha 2}$ radiation correction (SIEMENS software package EVA 5.0 rev.1). Rietveldt analysis was performed using the software program Topas R (Bruker, Karlsruhe, Germany). High temperature X-ray diffraction measurements were performed using a hot stage with a Pt tape as heater. A reaction between the Pt tape during the measurements and the sample was not observed. Heating time was 15 °C/min followed by a holding time of 60 min at constant temperature during the measurements.

RESULTS

An influence of the preparation method on the crystal structure of the compound was not observed (for further information, see also [34-36]. Both methods, mixed oxide route and combustion synthesis result in identical samples. For both routes of preparation, the cooling rates achieved were considered to be sufficiently high as to merely facilitate allotropic phase transitions upon cooling. The formation of low temperature compounds by long-distance diffusion was thus suppressed.

(1) Quasi Ternary Phase Diagram $LaGaO_3$-$SrGaO_{2.5}$-"$LaMgO_{2.5}$"

$LaGaO_3$ is a compound of the binary system La_2O_3-Ga_2O_3. When in nominal batch compositions (i.e., $La_{1-x}Sr_xGa_{1-y}Mg_yO_3$), La and Ga are substituted by Sr and Mg, respectively, the constitutional changes can be depicted in the quasi-ternary system $LaGaO_3$-$SrGaO_{2.5}$-"$LaMgO_{2.5}$", comprising the solid solution range of LSGM as well as several multi-phase regions. Within this system, the compound $SrGaO_{2.5}$ does indeed exist, whereas „$LaMgO_{2.5}$" merely represents the overall composition.

Figures 1-4 show isothermal sections of the system $LaGaO_3$-$SrGaO_{2.5}$-$LaMgO_{2.5}$ for low amounts of dopants (i.e., regions adjacent to single phase LSGM). At 1500°C, the homogeneity region of LSGM stretches far out into the quaternary. The solubility of either cation increases with the amount added of the complementary dopant. LSGM accomodates approximately up to 25 at% of Sr and up to 20 at% Mg, when both dopants are added simultaneously. With decreasing temperature, the single phase range diminishes of LSGM rapidly. It is particularly noteworthy that at a temperature as high as 1250°C the solubility of either dopant has decreased to less than 5 at% each.

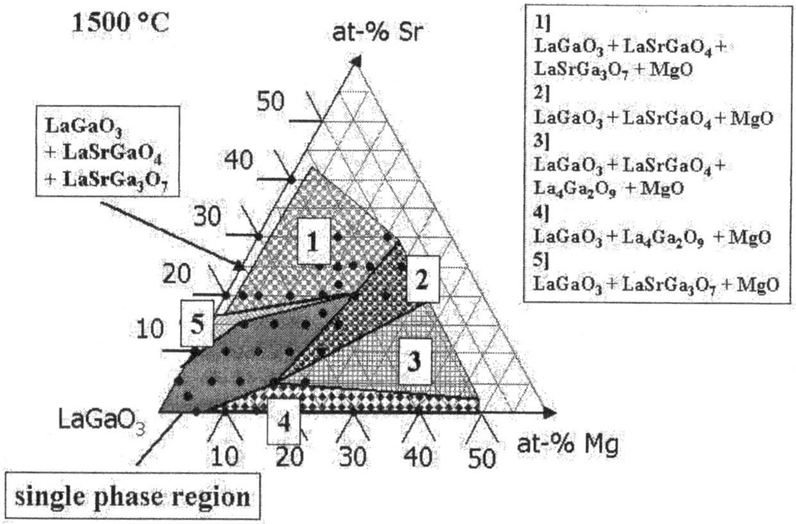

Figure 1: Isothermal sections of the system $LaO_{1.5}$-SrO-$GaO_{1.5}$-MgO at 1500 °C.

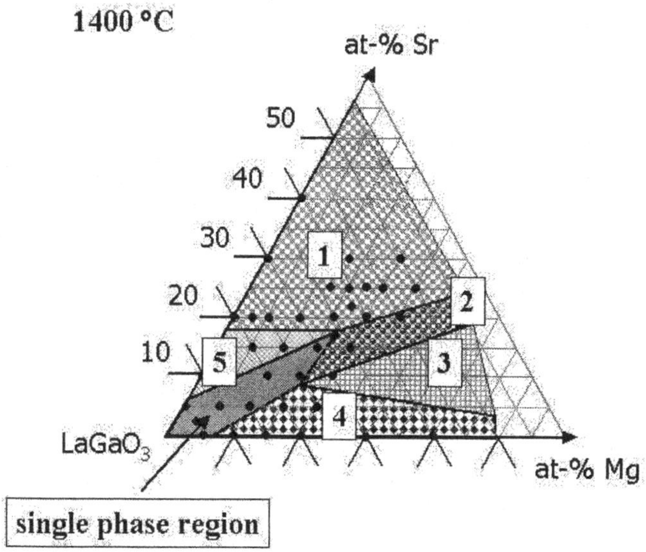

Figure 2: Isothermal sections of the system $LaO_{1.5}$-SrO-$GaO_{1.5}$-MgO at 1400 °C.

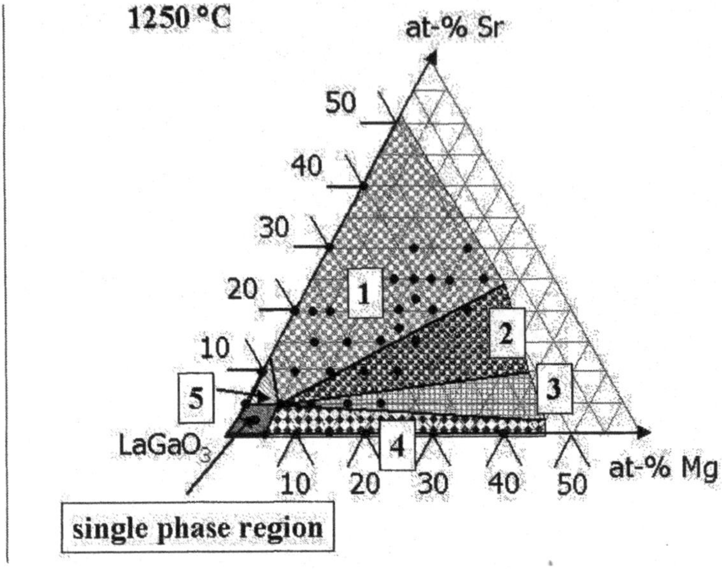

Figure 3: Isothermal sections of the system $LaO_{1.5}$-SrO-$GaO_{1.5}$-MgO at 1250 °C.

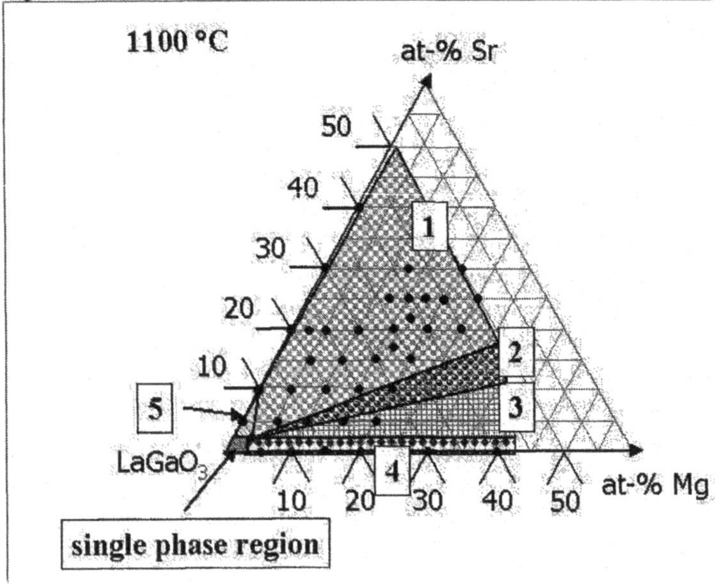

Figure 4: Isothermal sections of the system $LaO_{1.5}$-SrO-$GaO_{1.5}$-MgO at 1100 °C.

The single phase region of LSGM is surrounded by two and three phase regions. Depending on temperature and composition, one of these is in all cases LSGM (ICDD PDF 81-2300), being in equilibrium with MgO (ICDD PDF 45-946), $LaSrGaO_4$ (ICDD PDF 24-1208), $LaSrGa_3O_7$ (ICDD PDF 45-0637), and $La_4Ga_2O_9$ (ICDD PDF 37-1433).

In order to verify the observed temperature dependent decrease in solubility of LSGM for Sr and Mg, a single phase $La_{0.8}Sr_{0.2}Ga_{0.8}Mg_{0.2}O_3$ sample prepared at 1500°C was post-annealed at 1250°C for 48 hrs (Figure 5). The corresponding XRD patterns reveal the formation of significant amounts of $LaSrGaO_4$, $LaSrGa_3O_7$, and MgO during the post annealing treatment.

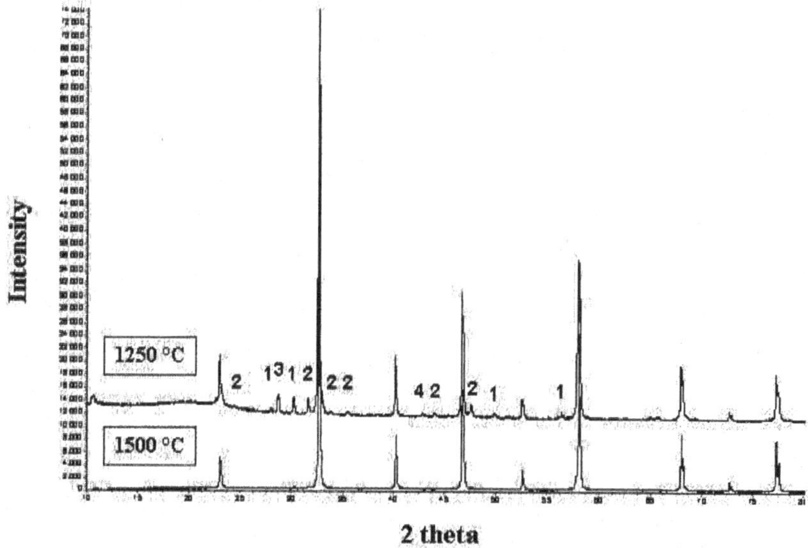

Figure 5: XRD patterns of LSGM ceramics ($La_{0.8}Sr_{0.2}Ga_{0.8}Mg_{0.2}O_3$) in the course of an anneal study; after sintering at 1500°C (bottom) and after subsequent annealing at 1250°C in air (top). 1: $LaSrGa_3O_7$; 2: $LaSrGaO_4$; 3: $La_4Ga_2O_9$; 4: MgO.

(2) Kinetic Study

A series of pelletized samples (precursor powders were prepared by combustion synthesis) was sintered at 1500°C for 50 h and subsequently annealed at 650°C for varying times of 50, 100, 500, 1000, 1500 h in an air atmosphere. The reduced anneal temperature was supposed to simulate operating conditions in RTSOFCs (under oxidizing atmospheres). In the bulk material, there was no detectable formation of secondary phases, as observed by SEM microstructural investigation and XRD (Figure 6). This finding indicates the long term (kinetic) stability of highly doped LSGM at reduced operating temperatures and in air atmosphere. However, this result does not necessarily imply the chemical stability in an electrochemical array, as interfacially induced formation of secondary phases between LSGM and variable electrode materials has been reported elsewhere [37].

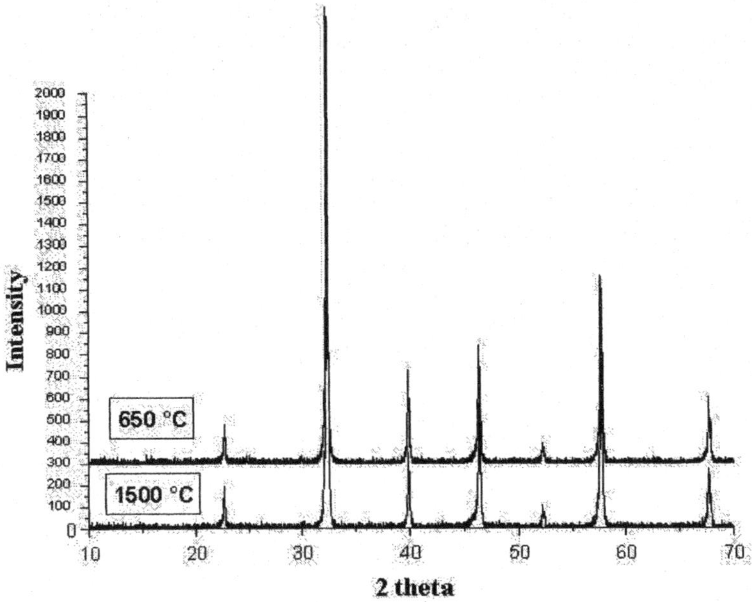

Figure 6: X-ray diffraction patterns of single phase $La_{0.8}Sr_{0.2}Ga_{0.8}Mg_{0.2}O_3$ prepared at 1500 °C and post annealed at 650°C.

Studies of the shrinking rate during sintering of LSGM powder prepared by a mixed oxide method show that significant shrinkage occurs at temperatures not below 1300 °C. Maximum densification was obtained at 1450 °C (Figure 7). For the preparation of dense ceramics of LSGM, sinter temperatures of about 1400 to 1500 °C have, therefore, to be considered.

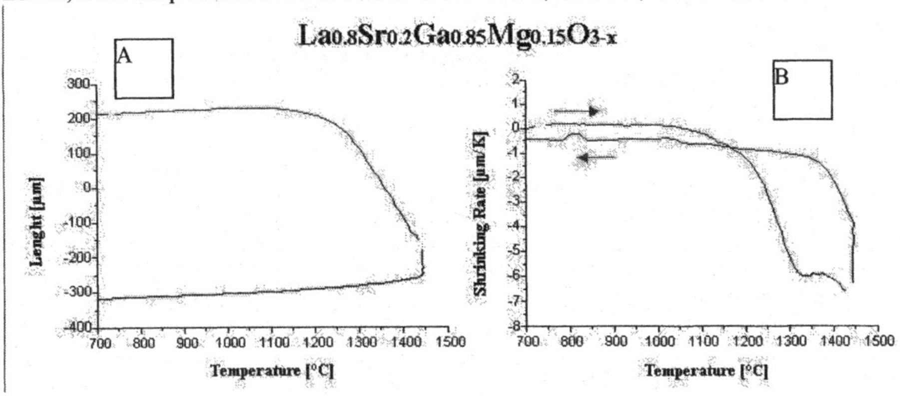

Figure 7: Densification (A) and shrinking rate (B) of LSGM.

(3) Crystal Strucuture

The structure of $La_{0.95}Sr_{0.05}GaO_3$ and $LaGa_{0.95}Mg_{0.05}O_3$ which represents almost the maximum solid solubility of either Sr or Mg in $LaGaO_3$ at 1500 °C in air [24] has been determined to be orthorhombic, space group Pbnm (see Table 1). These data correlate with pure $LaGaO_3$. LSGM with Sr and Mg contents of up to about 7.5 mol% each ($La_{0.85}Sr_{0.15}Ga_{0.85}Mg_{0.15}O_3$) was identified to be orthorhombic, space group Pbnm (see Table 1) (Figure 8). At a Sr and Mg content of 8.75 mol% each ($La_{0.825}Sr_{0.175}Ga_{0.825}Mg_{0.175}O_3$) the structure of the compound was determined to be rhombohedral, space group $R\bar{3}c$ with a = 5.5350 Å and c = 13.5580 Å (Figure 9). At maximum solid solubility of Sr and Mg in LSGM of almost 10 mol% ($La_{0.8}Sr_{0.2}Ga_{0.8}Mg_{0.2}O_3$ [4,7,12,14,24,25]) the structure was found to be cubic, space group $Pm\bar{3}m$, with a = 3.9109 Å (Figure 10).

Table 1: Crystallographic data of the prepared compounds.

Sample	a [Å]	b [Å]	c [Å]	space group
$La_{0.95}Sr_{0.05}GaO_3$	5.5246	5.4918	7.7749	Pbnm
$LaGa_{0.95}Mg_{0.05}O_3$	5.5258	5.4942	7.7780	Pbnm
$La_{0.9}Sr_{0.1}Ga_{0.9}Mg_{0.1}O_3$	5.5333	5.4965	7.7819	Pbnm
$La_{0.85}Sr_{0.15}Ga_{0.9}Mg_{0.1}O_3$	5.5281	5.5093	7.7770	Pbnm
$La_{0.85}Sr_{0.15}Ga_{0.85}Mg_{0.15}O_3$	5.5298	5.5068	7.8030	Pbnm
$La_{0.825}Sr_{0.175}Ga_{0.825}Mg_{0.175}O_3$	5.5350		13.5580	$R\bar{3}c$
$La_{0.8}Sr_{0.2}Ga_{0.8}Mg_{0.2}O_3$	3.9109			$Pm\bar{3}m$

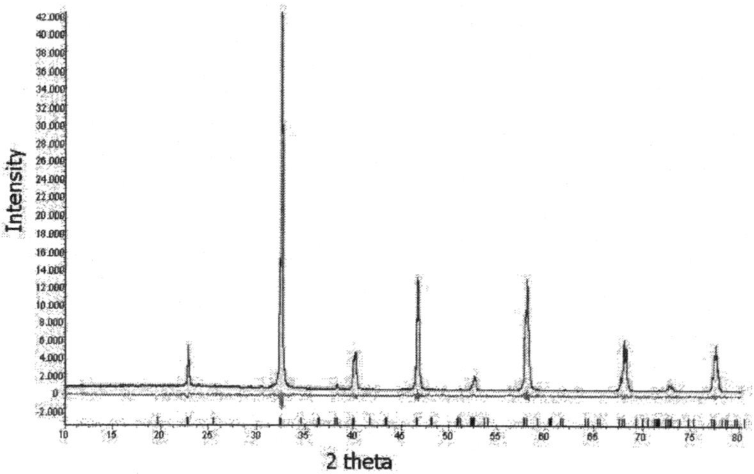

Figure 8: Rietveld analysis of $La_{0.85}Sr_{0.15}Ga_{0.85}Mg_{0.15}O_3$.

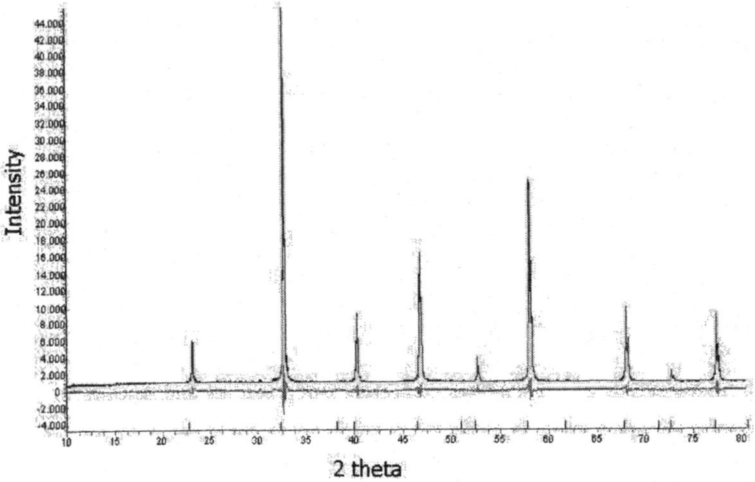

Figure 9: Rietveld analysis of $La_{0.825}Sr_{0.175}Ga_{0.825}Mg_{0.175}O_3$.

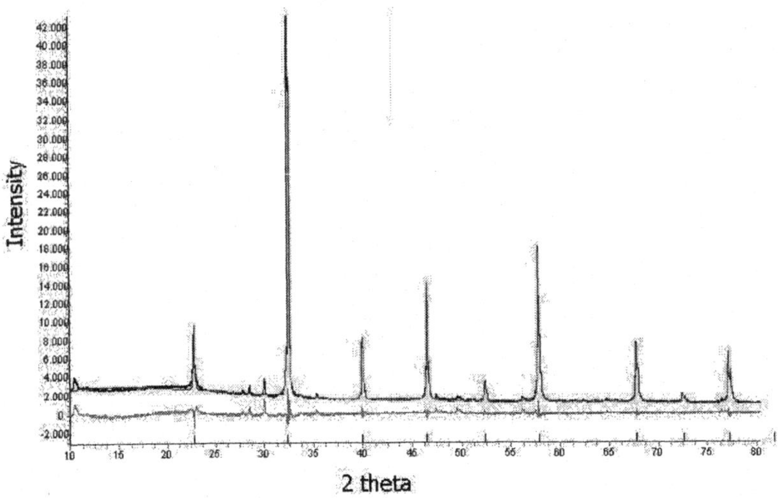

Figure 10: Rietveld analysis of $La_{0.8}Sr_{0.2}Ga_{0.8}Mg_{0.2}O_3$.

The changes of the crystal structure can clearly be observed in focusing the reflection at about 40 ° 2 theta. Figure 11 shows Rietveld refinements of that reflection. The refinements clearly show the change of the crystal structure of LSGM with increasing Sr and Mg content.

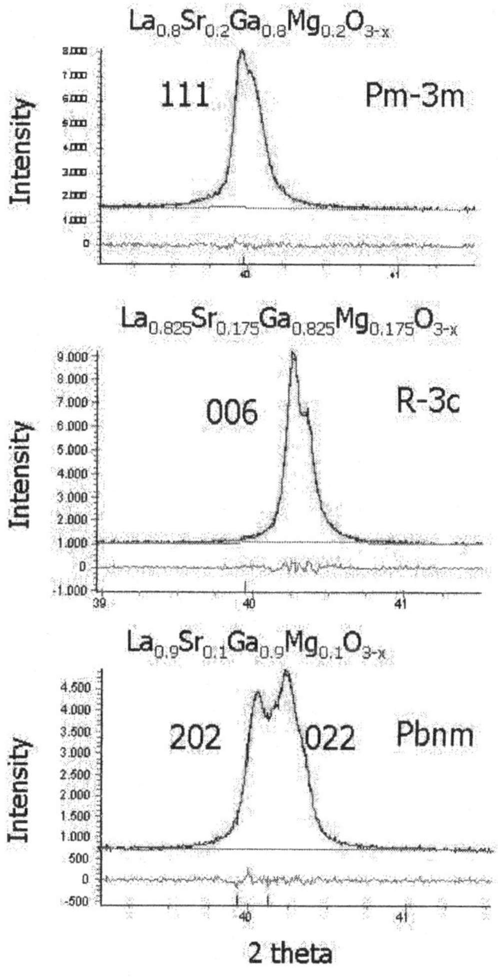

Figure 11: Rietveld refinements of the reflection at about 40 ° 2 theta clearly show the change of the crystal structure with increasing Sr and Mg content of LSGM.

XRD measurements at temperatures between 25 and 1000 °C of a sample having the composition $La_{0.85}Sr_{0.15}Ga_{0.9}Mg_{0.1}O_3$ show a phase transition from orthorhombic to cubic, space group $Pm\bar{3}m$, taking place at temperatures between 400 and 500 °C. This transition is accompanied by combined with a change in crystal density from $V_{ortho} = 6.8$ g/cm^3 to $V_{cubic} = 6.3$ g/cm^3. Thermal expansion coefficients were determined for all crystallographic directions and are given as followed: $\alpha_{a,\,ortho} = 10.81 \times 10^{-6}$ K^{-1}, $\alpha_{b,\,ortho} = 9.77 \times 10^{-6}$ K^{-1}, and $\alpha_{c,\,ortho} = 9.83 \times 10^{-6}$ K^{-1} of the orthorhombic modification between 25 and 400 °C as well as $\alpha_{cubic} = 13.668 \times 10^{-6}$ K^{-1} of the cubic modification between 500 and 1000 °C (Figure 12).

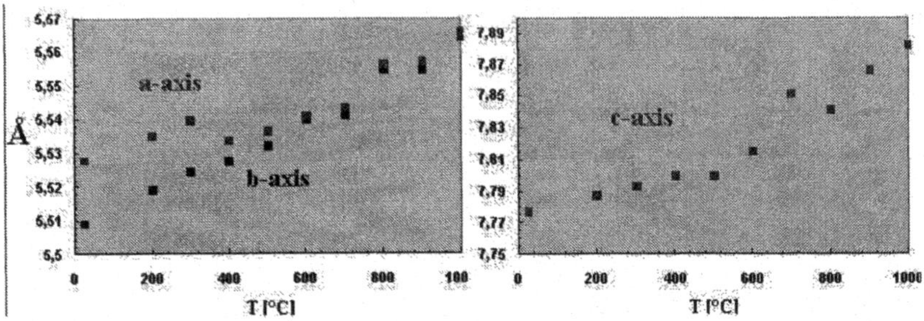

Figure 12: a, b, and c parameters of $La_{0.85}Sr_{0.15}Ga_{0.9}Mg_{0.1}O_3$ at a function of temperature.

Calculations of the individual atom positions based on the structural information obtained by Rietveld analysis reveal that the cubic modification exhibits a perfect 180 ° angle of the Ga-O-Ga links in all crystallographic directions. This finding indicate that the oxygen octrahedra are not tilted (Figure 13). In addition, the angle between the O-Ga-O which represent the degree of distortion of the oxygen octrahedra was calculated to be 90 °. This reveal that the oxygen octrahedra are not distorted. With decreasing Sr and Mg content the rhombohedral structure exists and the angles of the Ga-O-Ga links are calculated to be 162.37 ° and 180 ° and those of the O-Ga-O link 90 ° and 89.33 ° and 90.67 °, respectively (Figure 14). For the orthorhombic structure, Ga-O-Ga angles of 161.44 ° and 158.53 were calculated and those of the O-Ga-O links to be 89.84 ° and 90.16 °, respectively, as well as 89.43 ° and 90.57 °, respectively (Figure 15). Using the data for pure $LaGaO_3$ published by [6], Ga-O-Ga angles of 158.12 ° and 161.10 ° were calculated and those of the O-Ga-O links to be 89.85 ° and 90.15 °, respectively, as well as 89.53 ° and 90.47 °, respectively.

DISCUSSION

For a large number of LSGM samples discussed in the open literature, we were able to establish good agreement with respect to phase content and for a given temperature and atmosphere. Up to this study, the most complete set of thermodynamic data had been given by Huang *et al.*[5,6] The phase diagram and the extension of the single phase region of LSGM constructed from their data for a sintering temperature of 1470°C is in good agreement with our results for samples annealed at 1500°C. However, the location and extension of the multi phase regions which are surrounding LSGM differ which may be due to different sample preparation.

With respect to their constitutional character at a given composition and temperature, most of the LSGM samples studied by Djurado *et al.*[23] fit well into the phase diagram proposed in this paper.

In our opinion, however, a compound identified by these authors to be $Sr_3Ga_2O_6$ (ICDD PDF 24-1200) is in fact $LaSrGaO_4$, showing the 100% XRD reflex at a 2Θ angle of approximately 31.5°.

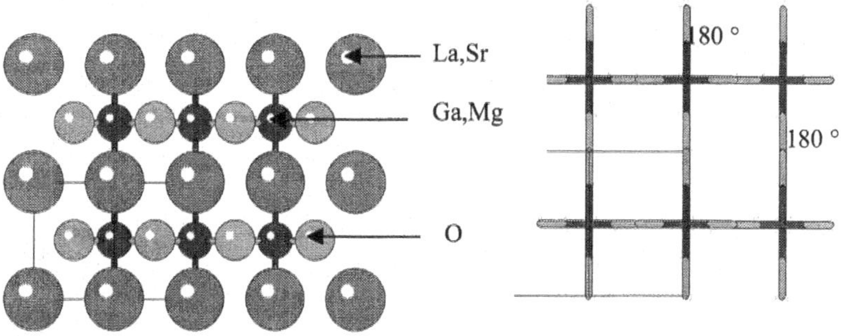

Figure 13: Sketch of the cubic structure. 110 plane is shown

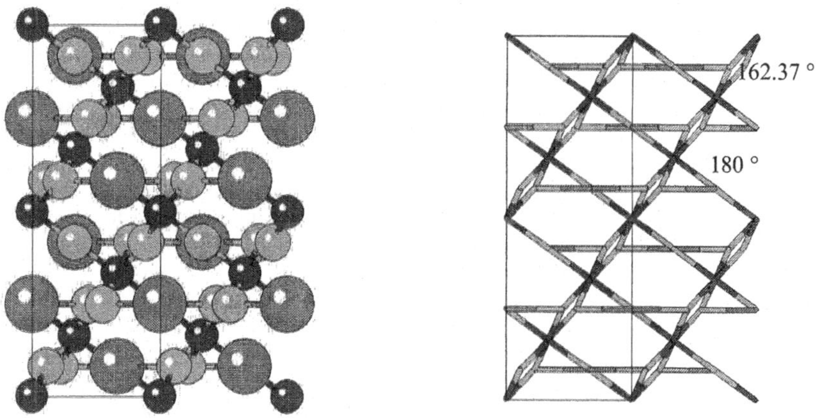

Figure 14: Sketch of the rhombohedral structure. 100 plane is shown.

In a recent patent[38], Goodenough and Huang report on the presence of a small amount of the impurity phase $LaSrGaO_4$ at low temperatures for a LSGM material with nominal composition $La_{0.9}Sr_{0.1}Ga_{0.8}Mg_{0.2}O_{2.85}$ sintered at 1470°C. According to our study, this phase evolves during sintering at high temperatures and is maintained at temperatures too low to facilitate solid state transport. In another study[1], the same authors also observed the formation of a small amount of $LaSrGaO_4$ on sintering at 1550°C and subsequent stack baking at 1125°C. Based on our observations, the occurrence of this impurity phase is bound to thermodynamics and its formation is indeed expected.

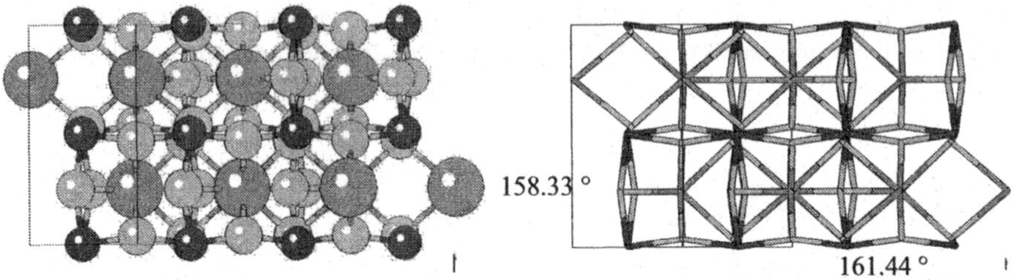

Figure 15: Sketch of the orthorhombic structure 110 plane is shown.

Wolfensteine *et al.*[18,19] sintered $La_{0.8}Sr_{0.2}Ga_{0.85}Mg_{0.15}O_{2.825}$ bulk samples at 1460°C for creep tests and state these to be single phase which is in agreement with the phase diagram for 1500°C presented in this work. Good agreement was also found for Sr solubility in $LaGaO_3$ of approximately 7.5 at.% with results by Ishihara *et al.*[12]. In contrast to this study, however, these authors identified $SrGaO_3$ and La_4SrO_7 as impurity phases which may be meta-stable impurity phases due to incomplete reaction of the starting material during the synthesis of LSGM [36].

One practical implication of this study is that, when a LSGM electrolyte of a composition tuned for high ionic conductivity (e.g., $La_{0.8}Sr_{0.2}Ga_{0.75}Mg_{0.25}O_{2.775}$) is manufactured at around 1470°C[4-6], one may predict decomposition reactions taking place upon cooling during fabrication as well as during operation in an electrochemical device. At operating temperatures of around 650°C, however, no formation of secondary phases was observed in this study.

A time dependent degradation of the conducting properties in a RTSOFC has recently been observed by Lybye *et al.*[39] and attributed to the formation of secondary phases. Taking into account the total conductivity of the major side-product $LaSrGa_3O_7$ to be three to four orders of magnitude lower than that of the LSGM electrolyte[38], this may be a straightforward assumption. As outlined above, however, we were not able in this study to identify the precipitation of impurity phases for an operating temperature of 650°C. It remains to be investigated in detail to what extend the possible formation of secondary phases may have an influence on the functional properties of a LSGM electrolyte under load. Structural and electro-chemical properties of melilite-type $LaSrGa_3O_7$ and its solid solutions are currently being under investigation by the present authors and will be published soon.

With respect to the here obtained structural data of LSGM and those data already published in literature the stability fields of the orthorhombic, rhombohedral, and cubic modification of LSGM are drawn in the quasi-ternary phase diagram $LaGaO_3$-$SrGaO_{2.5}$-$LaMgO_{2.5}$ (Figure 11). In good agreement with the data for $La_{0.9}Sr_{0.1}Ga_{0.8}Mg_{0.2}O_3$ published by [27] a temperature dependent phase transition from orthorhombic to cubic was observed. This observation indicate that the stability field of the cubic modification enlarges significantly with increasing temperature.

The observed difference in thermal expansion coefficients of both the orthorhombic and cubic modification may explain the differences of the already published values for polycrystalline samples of LSGM which vary between 10 and 12.47 × 10^{-6} K^{-1} depending on the temperature range of the corresponding measurements [40].

The symmetry of perovskites is mainly controlled by the edge sharing oxygen octrahedra of which distortion and tilt is strongly dependent on the size of the A and B cation. With decreasing symmetry, for example from cubic to orthorhombic, the angle between the octrahedra changes. Consequently, in case of $LaGaO_3$, the angle between the Ga-O-Ga links which represents the angle between neighboring oxygen octrahedra have to change with decreasing Sr and Mg content of the compound. These calculations reveal that for LSGM the observed phase transformations are controlled by tilt and distortion of the oxygen octrahedra the degree of which is depending on the Sr and Mg content.

CONCLUSION

Experimental phase equilibria studies within the quasi-quaternary system La_2O_3-SrO-Ga_2O_3-MgO were performed, aiming at the assessment of the temperature dependence of the solid solubility of Sr and Mg in LSGM. Powder and bulk samples were synthesized by conventional solid state processing as well as by a modified combustion synthesis and subsequently annealed at 1100°, 1250°, 1400°, and 1500°C in an air atmosphere. It was found that the single phase region of LSGM solid solutions decreases drastically with decreasing temperature. While at 1500°C, when both cations are added, an extended solubility was found for Sr (up to approximately 25 mol-%) and Mg (approximately 20 mol-%), the solubility for either cation at 1100 °C was established to be merely a few percent. The cubic allotropic form exists only at temperatures of 1400°C and above, and for high dopant amounts.

A kinetic study was carried out by long term annealing of electrolyte samples with a nominal composition of $La_{0.8}Sr_{0.2}Ga_{0.8}Mg_{0.2}O_{2.8}$. At a foreseeable RTSOFC operating temperature of 650°C, samples proved to stay thermochemically stable. No decomposition of side-phases such as $LaSrGa_3O_7$ was observed when annealed for up to 1500 h. In conclusion, it may be anticipated that solid electrolytes of this or a similar composition, when used in real devices, do not form secondary phases, when first prepared at elevated temperatures and eventually operated at temperatures as low as 650°C.

$LaGaO_3$ substituted with Sr and Mg exhibits a complex correlation between Sr and Mg content and crystal structure. At low Sr and Mg contents the compound is orthorhombic, whereas at higher Sr and Mg content a rhombohedral and cubic modification is found at room temperature. This phenomenon appears to be controlled by tilt and distortion of the oxygen octrahedra of the Perovskit structure as a function of the Sr and Mg content of the phase. With increasing temperature the stability field of the cubic modification extends in the direction of lower contents of Sr and Mg. The complex correlation between chemical composition of Sr and Mg substituted $LaGaO_3$ and its crystal structure is of great interest for the oxygen conducting properties of the compound, because beside vacancies in the oxygen sub lattice the crystal structure may also affects the oxygen conductivity. This fact is indicated by the phase transition from orthorhombic to cubic at about 500 °C which may correlate with the observed anomaly of the conducting properties at about the same temperature [4].

REFERENCES

[1]K. Huang, R. S. Tichy, and J. B. Goodenough, "Superior Perovskite Oxide-Ion Conductor; Strontium- and Magnesium-Doped $LaGaO_3$", *J. Electrochem. Soc.*, **144** [10] 3620-24 (1997)

[2]K. Huang, M. Feng, J. B. Goodenough, and M. Schmerling, "Characterisation of Sr-Doped $LaMnO_3$ and $LaCoO_3$ as Cathode Materials for a Doped $LaGaO_3$ Ceramic Fuel Cell", *J. Electrochem. Soc.*, **143** [11] 3630-36 (1996)

[3]K. Huang, M. Feng, and J. B. Goodenough, "Sol-Gel Synthesis of a New Oxide-Ion Conductor Sr- and Mg-Doped $LaGaO_3$ Perovskite", *J. Am. Ceram. Soc.*, **79** [4] 1100-104 (1996)

[4]K. Huang, R. S. Tichy, and J. B. Goodenough, "Superior Perovskite Oxide-Ion Conductor; Strontium- and Magnesium-doped $LaGaO_3$: I, Phase Relationships and Electrical Properties", *J. Am. Ceram. Soc.*, **81** [10] 2565-75 (1998)

[5]K. Huang, R. S. Tichy, and J. B. Goodenough, "Superior Perovskite Oxide-Ion Conductor; Strontium- and Magnesium-doped $LaGaO_3$: II, ac Impedance Spectroscopy", *J. Am. Ceram. Soc.*, **81** [10] 2576-80 (1998)

[6]K. Huang, R. S. Tichy, and J. B. Goodenough, "Superior Perovskite Oxide-Ion Conductor; Strontium- and Magnesium-doped $LaGaO_3$: III, Performance Tests of Single Ceramic Fuel Cells", *J. Am. Ceram. Soc.*, **81** [10] 2581-85 (1998)

[7]M. Feng and J. B. Goodenough, "A Superior Oxide-ion Electrolyte", *Eur. J. Solid State Inorg. Chem.*, **31**, 663-672 (1994)

[8]T. Ishihara, T. Shibayama, M. Honda, H. Nishiguchi, and Y. Takita, "Solid Oxide Fuel Cell Using Co-Doped $La(Sr)Ga(Mg)O_3$", *Chem. Commun.*, **9**, 1227-1228 (1999)

[9]T. Ishihara, M. Honda, T. Shibayama, H. Minami, H. Nishiguchi, and Y. Takita, "Intermediate Temperature Solid Oxide Fuell Cells Using a New $LaGaO_3$ Based Oxide Ion Conductor", *J. Electrochem. Soc.*, **145** [9] 3177-83 (1998)

[10]T. Ishihara, H. Matsuda, and Y. Takita, "Effects of Rare Earth Cations Doped for La Site on the Oxide Ionic Conductivity of $LaGaO_3$-based Perovskite-Type Oxide", *Solid State Ionics*, **79**, 147-151 (1995)

[11]T. Ishihara, J. A. Kilner, M. Honda, and Y. Takita, "Oxygen Surface Exchange and Diffusion in the New Perovskite Oxide Ion Conductor $LaGaO_3$", *J. Am. Ceram. Soc.*, **129**, 2747-2748 (1997)

[12]T. Ishihara, H. Matsuda, and Y. Takita, "Doped $LaGaO_3$ Perovskite Type Oxide as a New Oxide Ionic Conductor", *J. Am. Ceram. Soc.*, **116**, 3801-3803 (1994)

[13]T. Ishihara, M. Honda, T. Shibayama, H. Furutani, and Y. Takita, "An Intermediate Temperature Solid Oxide Fuel Cell Utilizing Superior Oxide Ion Conducting Electrolyte, Doubly Doped $LaGaO_3$ Perovskite", *Ionics*, **4**, 395-402 (1998)

[14]P. Huang and A. Petric, "Superior Oxygen Ion Conductivity of Lanthanum Gallate Doped with Strontium and Magnesium", *J. Electrochem. Soc.*, **143** [5] 1644-47 (1996)

[15]R. T. Baker, B. Gharbage, and F. M. B. Marques, "Ionic and Electronic Conduction in Fe and Cr Doped $(La,Sr)GaO_{3-\delta}$", *J. Electrochem. Soc.*, **144** [9] 3130-35 (1997)

[16]R. Maric, S. Ohara, T. Fukui, H. Yishida, M. Nishimura. T. Inagaki, and K. Miura, "Solid Oxide Fuell Cells with Doped Lanthanum Gallate Electrolyte and $LaSrCaO_3$ Cathode, and Ni-Samaria-Doped Ceria Cermet Anode", *J. Electrochem. Soc.*, **146** [6] 2006-2010 (1999)

[17]K. Yamaji, T. Horita, M. Ishikawa, N. Sakai, and H. Yokokawa, "Chemical Stability of the $La_{0,9}Sr_{0,1}Ga_{0,8}Mg_{0,2}O_{2,85}$ Electrolyte in a Reducing Atmosphere", *Solid State Ionics*, **121**, 217-224 (1999)

[18]J. Wolfenstine, "Rate-controlling Species for Creep of the Solid State Electrolyte: Doped Lanthanum Gallate", *Solid State Ionics*, **126**, 293-298 (1999)

[19]J. Wolfenstine, P. Huang, and A. Petric, "High-Temperature Mechanical Behavior of the Solid-State Electrolyte", *J. Electrochem. Soc.*, **147**, 1668-1670 (2000)

[20]L. Vasykechko, A. Matkovski, A. Suchicki, and I. Syvorotka, "Crystal Structure of $LaGaO_3$ and (La,Gd)GaO_3 Solid Solutions", *J. Alloy. Compd.*, **286**, 213-218 (1999)

[21]M. Hrovat, S. Bernik, J. Holc, and Z. Samardzija, "Subsolidus Phase Equilibria in the La_2O_3-GaO-NiO System", *J. Mater. Res.*, **14** [6] 2351-2354 (1999)

[22]A. Naoumidis, A. Ahmad-Khanlou, Z. Samardzifa, and D. Kolar, "Chemical Interactions and Diffusion on Interface Cathode/Electrolyte of SOFC)", *Fresenius J. Anal. Chem.*, **365**, 277-281 (1999)

[23]E. Djurado and M. Labeau, "Second Phases in doped Lanthanum Gallate Perovskites", *J. Eur. Ceram. Soc.*, **18**, 1396-1404 (1998)

[24]P. Majewski, M. Rozumek, H. Schluckwerder, and F. Aldinger, "Phase Diagram Studies in the Systems $LaO_{1.5}$-SrO-$GaO_{1.5}$, $LaO_{1.5}$-MgO-$GaO_{1.5}$, and SrO-MgO-$GaO_{1.5}$ at 1400°C in Air", *J. Am. Ceram. Soc.*, **84**, 1093-1096 (2001)

[25]P. Majewski, M. Rozumek, and F. Aldinger, "Phase diagram studies in the systems La_2O_3-SrO-MgO-Ga_2O_3 at 1350-1400 °C in air with emphasis on Sr and Mg substituted LaGa3", *J. Alloys Comp.*, **329**, 253-258 (2001)

[26]C.J. Howard, B.J. Kennedy, "The orthorhombic and rhombohedral phases of $LaGaO_3$ – a neutron powder diffraction study", J. Phys. Matter, **11** 3229-3236 (1999).

[27]M. Lerch, H. Boysen, T. Hansen, "High-temperature neutron scattering investigation of pure and doped lanthanum gallate", J. Phys. Chem. Solids, **62** 445-455 (2001).

[28]K. C. Patil, "Advanced Ceramics: Combustion Synthesis and Properties", *Bull. Mater. Sci.*, **16**, 533-541 (1993)

[29]J. W. Stevenson, T. R. Armstrong, D. E. McCready, L. R. Pederson, and W. J. Weber, "Processing and Electrical Properties of Alkaline Earth-Doped Lanthanum Gallate", *J. Electrochem. Soc.*, **144**, 3613-3620 (1997)

[30]J. W. Stevenson, T. R. Armstrong, L. R. Pederson, J. Li, C. A. Lewinsohn, and S. Baskaran, "Effect of A-site Cation Nonstoichiometry on the Properties of Doped Lanthanum Gallate", *Solid State Ionics*, **113-115**, 571-583 (1998)

[31]S. Baskaran, C. A. Lewinsohn, M. Quian, J. W. Stevenson, and T. R. Armstrong, "Mechanical Properties of Alkaline Earth-Doped Lanthanum Gallate", *J. Mater. Sci.,* **34**, 3913-3922 (1998)

[32]T. Mathews, J. R. Sellar, B. C. Muddle, and P. Manoravi, "Pulsed Laser Deposition of Doped Lanthanum Gallate and In Situ Analysis by Mass Spectrometry of the Laser Ablation Plume", *Chem. Mater.,* **12**, 917-922 (2000)

[33]S. R. Jain, K. C. Adiga, and V. R. P. Verneker, "A New Approach to Thermochemical Calculations of Condensed Fuel-Oxidizer Mixtures", *Combust. Flame*, **40**, 71 (1981)

[34]A. C. Tas, P. Majewski, and F. Aldinger, "Chemical Preparation of Sr- and Mg-Doped $LaGaO_3$ Solid Oxide Fuel Cell Powders" *J. Am. Ceram. Soc.*, **83**, 2954-2960 (2000)

[35]A.C. Tas, P. Majewski, and F. Aldinger, "Preparation of Sr- and Zn-doped $LaGaO_3$ by Precipitation in the Presence of Urea and/or Enzyme Urease", *J. Am. Ceram. Soc.*, **85**, 1414-1420 (2002)

[36]P. Majewski, M. Rozumek, C.A. Tas, F. Aldinger, Processing of $(La,Sr)(Ga,Mg)O_3$ Solid Electrolyte, *J. Electroceramics*, **8**, 65-73 (2002).

[37]M. Rozumek, P. Majewski, T. Maldener, and F. Aldinger, Study of the Solid State Reactions between $(La,Sr)(Ga,Mg)O_3$ and $(La,Sr)MnO_3$, $(La,Ca)CrO_3$, and Ni, *Mat.-wiss. und Werkstofftech.*, **33**, 348-354 (2002)

[38]US Patent Number: 6,004,688, Dec. 21, 1999

[39]D. Lybye, F. W. Poulsen, and M. Mogensen, "Conductivity of A- and B-site doped $LaAlO_3$, $LaGaO_3$, $LaScO_3$ and $LaInO_3$", *Solid State Ionics*, **128**, 91-103 (2000)

[40]H. Hayashi, M. Suzuki, H. Inaba, "Thermal expansion of Sr- and Mg-doped $LaGaO_3$", Solid State Ionics, **128** 131-139 (2000).

Phase Equilibria of the Lead-Magnesium-Titanate and Lead-Niobium-Titanate Systems at 1000°C and 1 atm

Robert Mangham* and William T. Petuskey*

Department of Chemistry, Arizona State University, Tempe, Arizona 85287-1604

Phase equilibria of the PbO-MgO-TiO_2 and PbO-Nb_2O_5-TiO_2 systems have been experimentally determined at 1000°C and 1 atm pressure. This work is part of an overall comprehensive thermal analysis of the ferroelectric relaxor system $Pb(Mg_{1/3}Nb_{2/3})O_3$-$PbTiO_3$. In the PbO-MgO-TiO_2 system, lead titanate equilibrates with each of the magnesium titanates, MgO and a titania saturated liquid lead oxide. The most significant feature in the PbO-Nb_2O_5-TiO_2 system is the extent of solid solution of the lead and oxygen-defect pyrochlore phase, $Pb_3(Nb_{1-x}Ti_x)_4O_{13}$, in which titanium dissolution is as much as $x = 0.394$ (or ~22.5 cation % Ti^{4+}).

I. Introduction

We undertook the general evaluation of the phase chemistry in the PbO-MgO-Nb_2O_5-TiO_2 system since it was needed as the basis for later investigations on the equilibrium vapor pressures of lead oxide over the perovskite solid solution $Pb(Mg_{1/3}Nb_{2/3})O_3$-$PbTiO_3$. Of the four external ternary diagrams, we have already reported on the PbO-MgO-Nb_2O_5 system at 1000°C, and the MgO-Nb_2O_5-TiO_2 system between 1000 and 1450°C.[1,2] We report here on the two other lead oxide-based ternaries, namely PbO-MgO-TiO_2 and PbO-Nb_2O_5-TiO_2. The phase equilibria of the quaternary system have also been evaluated and are presented in a thesis[3], and will be described in later publications.

In this investigation, we experimentally examine the phase compatibilities between binary compounds and compositional ranges for the PbO-MgO-TiO_2 and the PbO-Nb_2O_5-TiO_2 system at 1000°C and 1 atm O_2. Compositional data determined by electron probe microanalysis is also presented with emphasis on the phase equilibria surrounding the pyrochlore solid solution in the PbO-Nb_2O_5-TiO_2 system. Differential thermal analysis was used to resolve a discrepancy concerning of the existence of $Pb_3Nb_2O_8$ at 1000°C in the binary join between PbO and Nb_2O_5.

II. Experimental Procedures

(a) Sample preparation

Samples were prepared from the constituent oxides PbO, MgO, TiO_2, and Nb_2O_5. Yellow orthorhombic PbO (also known as β-PbO, or massicot, Alfa Aesar 99.99% metals basis) was first annealed at 650°C, for 10 hours in an MgO crucible and flowing oxygen. The product powder was lightly ground in an agate mortar and then stored in a desiccator until used. Rutile, TiO_2, (Aldrich Chemical Co 99.9 % metals basis <325 mesh), magnesium oxide, MgO, (Alfa Aesar 99.95% metals basis <325 mesh,) and niobium pentoxide, Nb_2O_5, (Aldrich Chemical Co 99.99% metal basis, <325 mesh) were each dried at 120°C in a drying oven prior to use. Thermogravimetric analysis showed that only the magnesium oxide retained small amounts of excess water. This was taken into account when preparing samples.

The lead niobates and lead titanates were made by combining stoichiometric quantities of the binary oxides, followed by grinding and mixing in either an agate mortar or tungsten carbide ball mill. The green powder was then pelletized or packed firmly into an MgO crucible, covered with an MgO lid, and, then placed in a box furnace in air, along side a type-R thermocouple. Platinum crucibles worked well only as long as when the atmosphere was continuously oxidizing. The samples were heated slowly and calcined at 800°C for about two hours. Afterwards, they were heated between 1000°C for the lead rich $Pb_5Nb_4O_{15}$ phase to 1100°C for niobium rich $PbNb_4O_{11}$ phase and allowed to anneal for 10 to 20 hours. Samples were rapidly cooled to room temperature typically within seconds.

The titanium niobates were prepared in a similar manner although at higher temperatures. Since vapor loss was not a problem in these cases, samples were openly calcined first at 1200°C to facilitate phase formation followed by a 1000°C equilibration anneal for much longer periods. The product pellets were typically very hard. Subsequent regrinding and annealing steps were avoided due to the extra attention given at the beginning to grinding, mixing, and longer initial anneals.

The magnesium titanates were prepared from MgO and TiO_2 powder mixtures. Both platinum and Al_2O_3 crucibles were used. The pressed pellets did not react with the Al_2O_3 due to minimal sample contact with crucible walls and the relatively low synthesis temperature 1000-1050°C. The samples were calcined for 20-40 hours in air. The magnesium rich phase Mg_2TiO_4 was marginally stable at 1000°C,[4] thus higher temperatures (1050°C) were used to ensure its formation.

(b) Phase equilibrium reactions

Phase equilibrium analysis was carried out using two general schemes. Initial phase compatibilities were determined by reacting binary compounds (e.g. $PbTiO_3$ + $TiNb_2O_7$) for extended anneals at 1000°C. Once the initial features of the phase diagram were established, compositional limits for the three-phase fields and any solid solutions were determined by reacting the constituent compound phases representing the equilibrium (e.g. TiO_2+$PbNb_2O_6$+ $TiNb_2O_7$) in the appropriate amounts. Both reaction schemes were used in complementary fashion, often with redundancy. To ensure equilibrium and no vapor loss of lead oxide, the phase compatibility anneals were carried out in sealed platinum capsules. The samples were hand-pressed into pellets (diameter ~3.5 mm) placed in platinum capsules (5mm x ~10mm), welded closed at both ends, and weighed before and after the anneals. Flowing oxygen outside the capsules protected against possible reaction on the inside and alloying of the capsule with lead. Some experiments were carried out for as long 200 hours or more. Each sample was quenched in water.

The second reaction scheme was of shorter calcining periods (40-60 hours). Pellets were made using a 1 cm hardened steel die with a Carver press for higher pressures (~300-450 MPa). Impurities scraped from the die walls upon pressing were removed by sanding the outer edges of the pellets with a diamond file. The pellets were then placed into MgO crucibles and buried in excess powder of the same composition. These were covered with an MgO lid and placed into a box furnace along side a type-R thermocouple in air. Samples were typically quenched in water. Both methods yielded the same x-ray diffraction patterns, but the hard-pellet method provided larger-grained microstructures, due to increased grain growth, which was more useful for microprobe analysis.

X-ray histograms were taken using a Siemens D5000 diffractometer equipped with a 7° wide position-sensitive-detector and Cu-K_α radiation. Compositional analyses were measured using a JEOL 8600 electron microprobe with a 40° takeoff angle. The samples were first imbedded in epoxy and polished down to a 0.25 μm finish with diamond paste. No etchants were used. Samples were coated with a 150 to 200 nm thick carbon layer. Composition standards for wavelength dispersive spectroscopy "WDS" were metallic niobium (Nb $L\alpha_1$) for niobium, metallic lead (Pb $M\alpha_1$) for lead, rutile TiO_2 for titanium (Ti $K\alpha_1$), MgO for magnesium (Mg $K\alpha_1$) and Al_2O_3 for oxygen (O $K\alpha_1$). A focused 2 μm beam running at 10 nA and 15 KeV accelerated voltage was used. Grains or single-phase polycrystalline domains greater than 4 μm were analyzed. Beam alignment and drift were checked regularly and recalibrated when necessary. The accuracy of the measured composition was assessed on known compounds in this system such as $MgTiO_3$, and $PbNb_2O_6$. Agreement was typically within 1-1.5 cation-%.

III. Results and Discussion

(a) PbO-MgO-TiO_2 Ternary Diagram

No ternary compounds exist in the PbO-MgO-TiO_2 system as shown in Fig 1. Several binary compounds coexist in equilibrium forming five clearly defined three-phase compatibility fields, as well as, a lead oxide rich liquid region. The MgO-TiO_2 binary is composed of three compounds.[5-7] Karroite, $MgTi_2O_5$, possesses the pseudobrookite structure, while geikielite, $MgTiO_3$, is a layered ilmenite, and quandilite, Mg_2TiO_4, is an inverse spinel. Wechsler and Navrotsky[4] showed that $MgTi_2O_5$ and Mg_2TiO_4 are stable only at high temperatures due to entropic contributions from cation disordering. Both phases dissociate at low temperatures to $MgTiO_3$ and TiO_2, and to $MgTiO_3$ and MgO, respectively. Wechsler and Von Dreele[8] showed complete ordering of Mg^{2+} and Ti^{4+} in $MgTiO_3$ on the two octahedral sites, and remains so up to

1400°C. In the higher temperature cubic Mg_2TiO_4 phase, the Mg^{2+} and Ti^{4+} ions are completely disordered on a single octahedral site. $MgTi_2O_5$ displays complete disordering of $Mg^{2+}$and Ti^{4+} at both low and high temperatures.

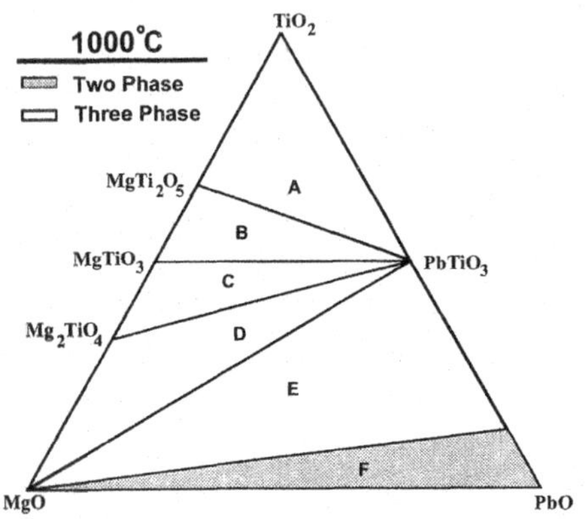

Fig. 1. Phase equilibrium diagram for the PbO-MgO-TiO_2 system at 1000°C and 1 atm. The light gray fields represent two phase domains, and the white fields represent three phase domains.

The PbO-TiO_2 binary system possesses only one compound, $PbTiO_3$.[9] Holman and Fulrath[10] estimated a small homogeneity range later confirmed by Popovic[11] and coworkers, where the mole fraction range of lead oxide was $0.495 < X_{PbO}^{in\ PbTiO_3} < 0.515$ at 1000°C. For clarity, we show $PbTiO_3$ to be simply a line compound. $PbTiO_3$ melts congruently at 1288°C.[12] No compounds are formed between PbO and MgO, nor is there any notable solubility of one to the other.

The electron microprobe analyses are given in Table 1 presented in cation %. Most notable in the PbO-MgO-TiO_2 ternary diagram is the absence of any ternary compound, and the absence of any solubility of Pb^{2+} into the magnesium titanates, and, conversely, no dissolution of Mg^{2+} into liquid lead oxide and lead titanate. This can be explained since the Pb^{2+} is much larger than Mg^{2+}, and therefore, do not fit comfortably in each other's lattice positions in the perovskite structure. On the other hand, rapidly quenched liquid lead oxide was found to crystallize in the high temperature orthorhombic PbO phase containing dissolved TiO_2 up to 13 cation % Ti^{4+}. This agrees with previous results attributed to Rase,[9] and Soh[12], as well as other DTA investigations by Stearns, *et al.*[13]

(b) PbO-Nb_2O_5-TiO_2 Ternary Diagram

The pyrochlore solid solution is the dominant feature of the PbO-Nb_2O_5–TiO_2 phase diagram, appearing in Figs 2 and 3. Numerous connecting two-phase fields radiate out in all directions. The only phases that do not equilibrate with the pyrochlore are $PbNb_4O_{11}$, $Ti_2Nb_{10}O_{29}$ and $TiNb_{24}O_{62}$. These are shown in Figs. 2 and 3. Table 2 lists the compositional determinations. One compositional limit of the pyrochlore phase, $Pb_{1.5}Nb_2O_{0.5}$, is without titanium. This corresponds to the defect pyrochlore formula that is sometimes expressed as $Pb_3Nb_4O_{13}$ (Lucas and Petuskey[1]). Hereafter, we refer to it as PNpy. Vacancies reside in equal quantities on both lead and oxygen sublattices. The other compositional extreme of the pyrochlore solid solution is $Pb_2(Nb_{.67}Ti_{.33})_2O_{6.67}$ reported by Bachelier, *et al.*[14]. We refer to it as PNTpy.

There are five distinct binary compounds along the PbO-Nb_2O_5 join at 1000°C, namely, $PbNb_4O_{11}$, $PbNb_2O_6$, $Pb_3Nb_4O_{13}$, $Pb_2Nb_2O_7$, and $Pb_5Nb_4O_{15}$, as reported by Roth.[15] Several independent studies have also revealed a series of homologous compounds between $Pb_5Nb_4O_{15}$ and $Pb_2Nb_2O_7$, having compositions $Pb_{2n+1}Nb_{2n-1}O_{7n-1}$, where n = 5,7,9, and 11[16-20], which can be described as layered pyrochlores with different stacking sequences. The phase chemistry of this series was not explicitly investigated in this study.

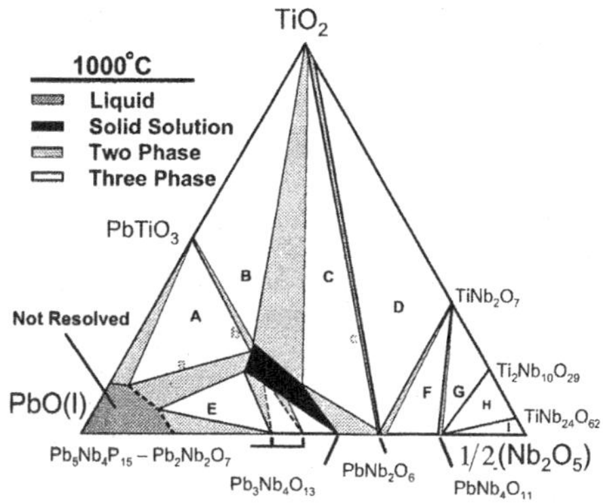

Fig. 2. Phase equilibrium diagram for the PbO-MgO-Nb_2O_5 system at 1000°C and 1 atm. The pyrochlore solid solution is shown in black, the two-phase domains is shown in light gray, and three phase domains are shown in white. Liquid lead oxide is shown to dissolve moderate amounts of TiO_2, but essentially not MgO.

With increasing temperature, the compound $Pb_3Nb_2O_8$ decomposes into $PbO(l)$ and $Pb_5Nb_4O_{15}(s)$ via a peritectic reaction. The precise temperature at which this happens is subject to some debate where it has been reported to be 1050 +/-5°C by Rivolier and co workers,[21] and 985°C by Roth.[15] Extended equilibrium anneals at 1000°C between PNTpy, PbO, and $Pb_5Nb_4O_{15}$ were used in this investigation to resolve this disagreement in thermal stability. As indicated in subset E (also see Table 2), $Pb_3Nb_2O_8$ in not thermodynamically stable at 1000°C.

The perovskite lead titanate, $PbTiO_3$ (PTpv), is the only binary compound in the PbO-TiO_2 system at 1000°C.[9,12] The TiO_2-Nb_2O_5 binary system has been the subject of several studies. Early high temperature studies by Roth[22] revealed two compounds; namely, $TiNb_2O_7$, and $TiNb_6O_{17}$. Only $TiNb_2O_7$ exists at 1000°C. Later investigations by Fedorov and coworkers[23] also reported the existence of two other niobium-rich phases, $Ti_2Nb_{10}O_{29}$ and $TiNb_{24}O_{62}$. $Ti_2Nb_{10}O_{29}$ exists in both orthorhombic and monoclinic crystal structures. However, both studies focused primarily at temperatures above 1000°C. Our investigations showed all three titanium-niobate phases ($TiNb_2O_7$, $Ti_2Nb_{10}O_{29}$, and $TiNb_{24}O_{62}$) exist at 1000°C, and each independently equilibrates with $PbNb_4O_{11}$ (see Fig 2.). The phase fields "F, G, H, and I" have been identified by x-ray diffraction after binary compatibility reactions and the compositions measured by EMPA of all of the phases in the three-field regions. The dissolution of Nb_2O_5 into rutile TiO_2 at 1000°C was reported at ~4.5 cation % by Dirstine and Rosa.[24] These results are in good agreement with our previous work on the MgO-Nb_2O_5-TiO_2 system.[2] Compositional analysis of phase field "C" and "D" revealed approximately 2 and 5.5 cation % Nb^{5+} in TiO_2, respectively.

The ternary pyrochlore (PNTpy) coexists with every other compound in this system except, Nb_2O_5, and the niobium-rich phase $PbNb_4O_{11}$. PNTpy coexists in broad two-phase regions with PbO, TiO_2, $PbNb_2O_6$, and, to a lesser extent, PTpv. Samples of PNTpy ($Pb_2(Nb_{.67}Ti_{.33})_2O_{6.67}$) were independently reacted with each of the following binary pyrochlores; $Pb_3Nb_4O_{13}$, $Pb_2Nb_2O_7$, $Pb_{2.31}Nb_2O_{7.31}$, and $Pb_5Nb_4O_{15}$. In all cases, binary compatibilities were established for each of the pyrochlores equilibrating with PNTpy. In Fig. 2, the series of "layered" pyrochlores between $Pb_2Nb_2O_7$ and $Pb_5Nb_4O_{15}$ is represented by the dashed lines leaving the details of this domain for future work. However, the width of the binary joins PNTpy-$Pb_2Nb_2O_7$, and PNTpy-$Pb_5Nb_4O_{15}$, were estimated from x-ray diffraction shifts in PNTpy and the boundaries of the neighboring three-phase fields. Several reactions with varying compositions along the PNTpy-$Pb_3Nb_4O_{13}$ join indicate a continuous solid solution between the two. For further information concerning pyrochlores, Subramanian, et, reported a detailed review. *al.*[25]

Further dissolution of Ti^{4+} into the pyrochlore was observed when in the presence of PTpv, PbO, and TiO_2. As shown in Fig's 2, 3 and Table 2, the pyrochlore solid solution extends to ~22.5 cation % Ti^{4+} when in equilibrium with TiO_2 and PTpv (domain "B") and ~20.5 cation % Ti^{4+} when equilibrating with PbO and PTpv (domain "A"). The pyrochlore compositions are accordingly $Pb_{3.84}(Nb_{0.56}Ti_{0.44})_4O_{12.956}$ and $Pb_{2.08}(Nb_{.56}Ti_{.40})_{1.92}O_{6.30}$. The composition of the pyrochlore in domain "C" at ~12.5 cation % Ti^{4+} corresponds to $Pb_{3.13}(Nb_{0.78}Ti_{0.23})_4O_{12.68}$. These experiments were repeated with the same results.

The compositional width of the PNTpy-TiO_2 binary equilibrium was determined by compositional limits imposed by the two adjacent ternary phase domains PNTpy-PTpv-TiO_2, (domain "B") and PNTpy-TiO_2-$PbNb_2O_6$ (domain "C"). A representative electron microprobe image, see Fig 7, shows the three-phase equilibrium phase field PNTpy-TiO_2-$PbNb_2O_6$ (domain "C"). The backscatter electron image (a), on the left clearly shows the coexistence of three phases. The dark/black phase is rutile TiO_2, with ~2.5 cation % Nb^{5+}. Microprobe analysis across the width of the TiO_2 grains revealed a constant Nb^{5+} concentration throughout, indicating that

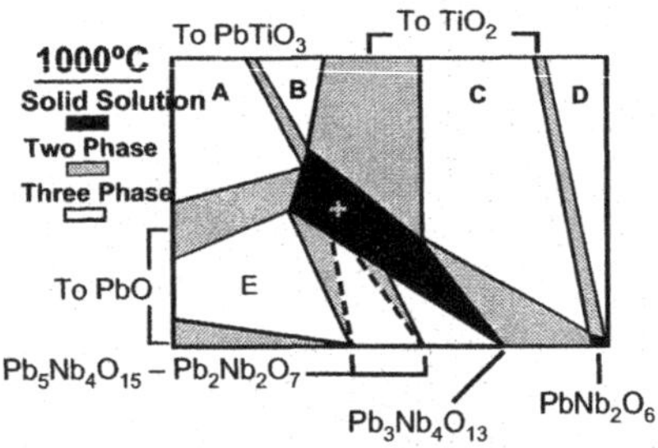

Fig. 3. Magnified view of the cubic pyrochlore solid solution domain in the PbO-Nb_2O_5-TiO_2 system at 1000°C and 1 atm. The grey "+" in the black pyrochlore domain represents the composition $Pb_2(Nb_{.66}Ti_{.33})_2O_{6.67}$. The measured cation compositions for the pyrochlore in the various three phase domains are given in Table 2.

thorough homogenization had been achieved. The light gray phase is $PbNb_2O_6$ (~1.5 cation % Ti^{4+}), and the darker gray phase is PNTpy (~12.5 cation % Ti^{4+}), of the composition $Pb_{3.13}(Nb_{0.78}Ti_{0.23})_4O_{12.68}$.

The domain labeled "F" in Fig. 2 was established and confirmed by multiple experiments, by reacting constituent oxides, i.e., by reacting combinations of ternary compounds, and by thermally equilibrating $PbNb_2O_6$ and $TiNb_2O_7$. The other niobium-rich domains labeled "G", "H", "I" were established by reacting each of the titanium niobates individually with $PbNb_2O_6$ and $PbNb_4O_{11}$ for extended periods in welded platinum capsules. Equilibrium reactions of several compositions with the three-phase domains "C" and "D" and analysis of the binary region between PNTpy and $PbNb_2O_6$, indicates $PbNb_2O_6$ dissolves up to 1.5 cation % Ti^{4+}.

The extent of the liquidus for the lead oxide solution was determined by microprobe analysis on water quenched samples after extended anneals at 1000°C in welded platinum capsules under flowing oxygen atmospheres. In the PbO-PTpv-PNTpy ternary region, liquid PbO dissolves 12.3(4) cation % Ti^{4+} and 4.6(2) % Nb^{5+}. However, grain structures in the PbO-$Pb_5Nb_4O_{15}$-PNTpy subset were too small for reliable microprobe analysis. Therefore the PbO liquidus is estimated and represented in the diagram by dashed lines. Confirmation of the compatibility between PNTpy and liquid PbO was determined by x-ray diffraction. The width of the binary join "a" was estimated from the two adjacent ternary fields, "A", and "E". The phase width of PbO-PTpv and PbO-$Pb_5Nb_4O_{15}$ binary domains were estimated from the adjacent ternary phase fields "A", and "E", respectively.

(c) Comments on the pyrochlore solid solution range

The PbO-Nb_2O_5-TiO_2 equilibrium diagram exhibits significant similarities to that previously reported for the PbO-MgO-Nb_2O_5 system at 1000°C, particularly with respect to the pyrochlore phase. The lead and oxygen defect pyrochlore PNpy ($Pb_3Nb_4O_{13}$) shows an extended solid solution range with the substitution up to 22.5 cation % Ti^{4+}. This is in contrast to the PMNpy in the PbO-MgO-Nb_2O_5 system where only ~7.5 cation % Mg^{2+} could be dissolved. The substitutional mechanisms of Mg^{2+} and Ti^{4+} into the pyrochlore are both presumed to involve the octahedral Nb-site although charge compensation imposes different numerical substitutional restrictions. For PMN pyrochlores, adding 3/2 Pb2+ maintains the charge balance for each Mg^{2+} added to a maximum value of 7.5 cation % Mg^{2+}. On the other hand, for in the PNTpy system, two mechanisms are exhibited. TiO_2 substitution up to 12.5 cation % Ti^{4+} are charged balanced by the creations of oxygen vacancies (1/2 $V_O^{\bullet\bullet}$), with essentially no change in lead oxide compositions. From 12.5 to 22.5 cation % added Ti^{4+}, the oxygen composition remains constant, while the composition of Pb^{2+} increases as the Pb^{2+} reside on lead vacancies sites with the ratio 1/2 Pb^{2+} for each Ti^{4+} added.

IV. Summary

The phase equilibrium diagrams for the $PbO-MgO-TiO_2$ and $PbO-Nb_2O_5-TiO_2$ systems were determined at 1000°C in oxygen atmospheres. The vaporization loss of PbO during synthesis and equilibrium reactions was minimized using encapsulation or buffering procedures. Extended runs of days to weeks were carried out in welded platinum capsules with flowing oxygen, whereas shorter runs (2-3days) could be carried out with pellets buried in excess powders of the same compositions inside counter nested MgO crucibles. Compositions and phase purity were evaluated using x-ray powder diffraction and electron microprobe analysis.

The $PbO-MgO-TiO_2$ phase diagram is a rather simple, involving four ternary compounds of fixed stoichiometry, a liquid oxide of modest composition range, and no quaternary compounds within the compositional field. All compounds, including the binary oxide components equilibrate with $PbTiO_3$, thus indicating the relative stability of the perovskite phase. The liquid lead oxide is capable of dissolving up to ~13 cation % TiO_2 in equilibrium, whereas there is no detectable solubility of MgO. As such, MgO is compositionally compatible and non-invasive when contact with the perovskite phase and liquid.

The $PbO-Nb_2O_5-TiO_2$ equilibrium phase diagram is similar in the sense that there is one phase that dominates most of the equilibrium relationships, however, it is the pyrochlore phase that is dominant, rather than the perovskite phase. The pyrochlore is shown to possess an extensive solid solution range between $Pb_{1.5}Nb_2O_{6.5}$ to $Pb_2(Nb_{.67}\ Ti_{.33})_2O_{6.67}$ and beyond to higher contents of TiO_2. The pyrochlore phase dissolves up to 22.5 cation % Ti^{4+}, which is much greater than magnesium dissolving into the same structure (i.e., ~7.5 cation % Mg^{2+}). The difference is likely related to the different cation sizes to substitute on the octahedral niobium site and the different charge compensation mechanisms involved. Like in the $PbO-MgO-Nb_2O_5$ system, the pyrochlore phase equilibrates with nearly every other compound of the system, with the exception of the most niobium-rich phases. Comparatively, conditions for thermodynamic stability of the perovskite phases, $PbTiO_3$ and $Pb(Mg_{1/3}Nb_{2/3})O_3$ within each of their respectively compositional fields, seem to be much narrower than for the pyrochlore phases. This explains in part the ease with which the pyrochlore appears during synthesis and high temperature treatment of the perovskites. In short, the volatilization of PbO from compositions intended for perovskite formation will invariably lead to the formation of pyrochlores.

Acknowledgements

Financial support from the Air Force Office of Scientific Research under Grant No. AFOSR F49620-02-1-0144 is gratefully acknowledged.

*Member of the American Ceramic Society

References

[1]P. Lucas, W.T. Petuskey "Phase Equilibria in the Lead-Magnesium-Niobium-Oxygen System at 1000°C," *J. Am. Ceram. Soc.*, **84** [9] 2150-52 (2001).

[2]Robert I. Mangham, "The High-Temperature Phase Chemistry and Thermochemistry of the Lead Magnesium Niobium Titanium Oxide System," Ph.D. Thesis, Arizona State University, Tempe, Arizona (2003).

[3]R. Mangham, W.T. Petuskey "High Temperature Phase Equilibria of the Magnesium-Niobium-Titanate System," *J. Am. Ceram. Soc.*, **84** [9] 2150-52 (2004).

[4]B. A. Wechsler and A. Navrotsky ,"Thermodynamics and structural Chemistry of Compounds in the System MgO-TiO_2" *J Solid. State. Chem.,* **55**, 165-80 (1984).

[5]T. F. W. Barth and E. Posnjak., "Spinel structures: with and without variant atom equipoints," *Z. Kristallogr. Teil A.*, **82** 325-41 (1932).

[6]Pauling., "The crystal structure of pseudobrookite," *Z. Kristallogr* **73**, 97-112. (1930).

[7]E. Posnjak and T. F. W. Barth., "Notes on some structures of the ilmenites," *Z. Kristallogr. Teil. A.,* **88** 271-80 (1934).

[8]B. A. Wechsler. R.B. Von Dreele., "Structure Refinements of Mg_2TiO_4, $MgTiO_3$, $MgTi_2O_5$ by Time-of-Flight Neutron Powder Diffraction", *Acta. Cryst.,* **B45** 542-49 (1989).

[9] D. E. Rase. Cited in B Jaffe, W.R. Crook and H Jaffey, Piezoeelectrics Ceramics, pp 117, Academic Press, London new York (1971).

[10]R. L. Holman, R. M. Fulrath., "Intrensic Nonstoichiometry in the Lead Zirconate-Lead Titinate System determined by Knudsen Effusion," *J. App. Phys.*, **44** 5227-36 (1973).

[11]A. Popovi_, B. Mali_ and LBencze., "Knudsen Cell Mass Spectromrttic Investigation of the PbO-ZrO2-TiO2 System" Rapid. Comm. Mass. Spectrom.,13 1129-37 (1999).

[12] J. R. Soh, H. M. Lee, H.S. Kwan., "Thermodynamic Evaluation and Phase Diagram of PbO-TiO_2 System." *Calphad.*, **18** [3] 237-44, (1984).

[13]L. Stearns and W. Petuskey, unpublished (2002).

[14]J. Bachelier, M. Hervieu, M. Quemeneur., "Contribution a L'Etude du Compose$Pb_{2-_}M_2O_{7-_}$ (M= Nb, Ti) de Type Phchlore," *E. Bull. Soc. Chim.* Fr., **9-10** 2593-97 (1973).

[15]R. S. Roth, "Phase Equilibrium Relations in the Binary System Lead Oxide-Niobium Pentoxide," *J.Res. Natl. Bur. Stand.* (U.S.), **62** [1] 27-38 (1959).

[16]M. C. Saine, G. Schiffmacher, M.Gasperin, H. Brusset, "Etude par Microscopic Electronique et Rayon X d'une Series de Niobates de Plomb Derives de la Structure Pyrochlore," *Rev. Chim. Min.*, **16** 597-06 (1979).

[17]M. C. Saine, M. Gasperin, H. Brusset, " Etude Cristallochimique de Cing Phases de Niobates de Plomb Derivant de la Structure Pyrochlore," *Rev. Chim. Min.*, **18** 587-92 (1981).

[18]C. Caranoni, F. Thuriest, C. Leroux, G. Nihoul, "Superstructures of Lead Niobium Oxides Diffraction Experiments and Modeling," *Philos. Mag. A,* **69** [3] 633-43 (1994).

[19]C. Leroux, H. Tatarenko, G. Nihoul., " High-resolution Electron Microscopy and modeling of Homologous Series in Nonstoichiometric Lead-Niobium Oxides," *Phys Rev. B*, **53** [18] 11993-05 (1996).

[20]C. Leroux, T. Badeche, G. Nihoul, O. Richard, G. van Tendeloo, " A Homologous Series $Pb_{2n+1}Nb_{2n-1}O_{7n-1}$ Studied by Electron Microscopy," *Eur. Phys. J. Appl. Phys.*, **7** [1] 33-40 (1999).

[21]J. L. Rivolier, M. Ferriol, M. T. Cohen-Adad., "The binary system $PbO-Nb_2O_5$,' *Eur. J. Solid. State. Inorg. Chem.*, t. **32** 251-62 (1995).
[22]R.S. Roth, L.W. Coughanour, "Phase equilibria relations in the systems titania-niobia and zirconia-niobia," *J. Res. Nat. Bur. Stand.*, 55 [4] 209-13 (1955).
[23]N.F. Fedorov et al., "The Nb_2O_5-TiO_2 System," *Russ.J.Inorg. Chem.*, **34** [5] 1316-19 (1989).
[24]R. T. Dirstine, and C. J. Rosa,"Defect Structure and Related Thermodynamic properties of nonstoichiometric Rutile (TiO_{2-x}) and Nb_2O_5 Doped Rutile," *Z Metallkund* **70** [6] 372-78 (1979).
[25]Subram, Aravamudan, Rao., "Oxide Pyrochlores- A Review," *Prog. Solid. St. Chem.*, 15 55-143 (1983).
[26]N. Wakiya, B.H. Kim, K. Shinozaki, and n. Mizutani., "Compositional Range of Cubic Pyrochlore Type Compound in Lead-Magnesiu-Niobium-Oxygen System," *J Ceram Soc. Jpn.*, **102** 615-17 (1994).

Table I. Cation Composition for the PbO-MgO-TiO_2 System

		% - Cation Composition ‡		
Domain	Phase	Pb	Mg	Ti
A	$PbTiO_3$	49.6 (4)*	0	50.3 (4)
	TiO_2	0	0	100.0
	$MgTi_2O_5$	0	33.9 (8)	66.0 (8)
		0	0	0
B	$PbTiO_3$	49.6 (04)	0	50.3 (04)
	$MgTi_2O_5$	0	34.2 (2)	65.5 (2)
	$MgTiO_3$	0	50.5 (2)	49.4 (3)
		0	0	0
C	$PbTiO_3$	49.9 (04)	0	50.0 (06)
	$MgTiO_3$	0	51.0 (2)	48.9 (2)
	Mg_2TiO_4	0	67.9 (1)	32.0 (1)
		0	0	0
D	$PbTiO_3$	50.1 (5)	0	49.8 (4)
	Mg_2TiO_4	0	66.8 (1)	32.2 (1)
	MgO	0	100	0
		0	0	0
E	$PbTiO_3$	49.8 (3)	0	50.1 (3)
	MgO	0	100	0
	PbO (*l*)	87.7 (7)	0	12.2 (8)
		0	0	0
F	MgO	0	100	0
	PbO (*l*)	86.9 (5)	0	13.1 (5)

‡ Determined by electron microprobe WDS
* "( )"refers to standard deviations

Table II. Cation Composition for the PbO-Nb_2O_5-TiO_2 System

		% - Cation Composition ‡		
domain	phase	Pb	Nb	Ti
A	PNTpy	50.6 (3)*	28.7 (5)	20.6 (8)
	$PbTiO_3$	50.0 (4)		49.3 (3)
	PbO	82.9 (8)	4.6 (2)	12.3 (4)
B	PNTpy	48.9 (5)	28.6 (7)	22.4 (5)
	$PbTiO_3$	50.2 (07)		49.7 (7)
	TiO_2	1.5 (7)		98.5 (4)
C	PNTpy	43.8 (2)	43.4 (4)	12.6 (2)
	TiO_2		2.4 (4)	97.5 (4)
	$PbNb_2O_6$	33.1 (6)	65.2 (7)	1.6 (6)
D	TiO_2		5.6 (8)	94.4 (7)
	$PbNb_2O_6$	32.9 (2)	65.5 (5)	1.4 (9)
	$TiNb_2O_7$		65.5 (1)	33.5 (1)
E	PNTpy	54.8 (6)	29.2 (5)	15.9 (7)
	$Pb_5Nb_4O_{15}$	54.3 (8)	43.1 (2)	2.5 (7)
	PbO (*l*)†			
a	PNTpy	49.9 (6)	32.2 (3)	17.7 (5)
	PbO (*l*)†			
b	PNTpy	49.6 (4)	29.3 (7)	21.1 (3)
	$PbTiO_3$	50.3 (2)		49.8 (5)
c	TiO_2		1.8 (5)	98.1 (4)
	$PbNb_2O_6$	33.8 (4)	64.9 (5)	1.2 (3)

‡ Determined by electron microprobe WDS
* "( )" refers to standard deviations
† Grain size too small for microprobe analysis

FERROELECTRIC BEHAVIOR IN THE AURIVILLIUS PHASE $Bi_3Ti_{1.5}W_{0.5}O_9$

Neil C. Hyatt*
Department of Engineering Materials,
The University of Sheffield,
Mappin Street,
Sheffield, S1 3JD. UK.

Kevin S. Knight
ISIS Pulsed Neutron and Muon Source,
Rutherford Appleton Laboratory,
Chilton,
Didcot,
Oxford, OX11 0QX. UK.

ABSTRACT

The ferroelectric phase transition in the Aurivillius phase $Bi_3Ti_{1.5}W_{0.5}O_9$ was studied by high resolution neutron powder diffraction. $Bi_3Ti_{1.5}W_{0.5}O_9$ adopts the polar orthorhombic space group $A2_1am$ at temperatures below the ferroelectric Curie temperature, $T_c = 735°C$. A direct transition to the non-polar tetragonal space group I4/mmm is observed on passing through T_c, with no evidence of an intermediate non-polar orthorhombic Amam phase. Above 700°C, $Bi_3Ti_{1.5}W_{0.5}O_9$ is metrically "tetragonal" (i.e. $a = b$, within experimental precision), however, the presence of weak supercell reflections associated with space group $A2_1am$ indicate that the material remains crystallographically orthorhombic below T_c.

INTRODUCTION

Ferroelectric layered perovskites of the Aurivillius family are of current interest for application in non-volatile Ferroelectric Random Access Memory (FRAM) devices [1-4]. The family of Aurivillius oxides, of general formula $Bi_2A_{n-1}B_nO_{3n+3}$, may be considered, in structural terms, as an ordered intergrowth of $[Bi_2O_2]^{2+}$ fluorite type and $[A_{n-1}B_nO_{3n+1}]^{2-}$ perovskite type layers (where A is a large 12 co-ordinate cation and B a small 6 co-ordinate cation with a d^0 electron configuration) [5, 6]. The transition between the low temperature ferroelectric and high temperature paraelectric phase(s) in this family of materials is, at present, the subject of considerable interest and debate. In double-layered Aurivillius phases, with $n = 2$, the ferroelectric phase adopts the polar orthorhombic space group $A2_1am$ and, above the ferroelectric Curie temperature, T_c, transforms to the archetype parent phase in space group I4/mmm [7-10]. Recent neutron and synchrotron X-ray powder diffraction studies have indicated that this transition may proceed *via* an intermediate paraelectric but orthorhombic phase in space group Amam [7-9]; evidence for such an intermediate phase was also observed in high temperature electron diffraction studies of the higher $n = 3$, 4 and 5 homologues [11]. For $n = 2$ Aurivillius phases, the intermediate paraelectric orthorhombic phase corresponds to the loss of a displacive mode and a tilt mode, relative to the ferroelectric phase, above T_c. The remaining tilt mode is retained and lost on the transformation to the archetype I4/mmm phase. This sequence of phase transitions has been shown to be continuous and is characteristic of the majority of $n = 2$ Aurivillius phases so far studied, *e.g.* $Bi_2SrTa_2O_9$, $Bi_{2.1}Sr_{0.85}Ta_2O_9$, $Bi_2PbTa_2O_9$ and $Bi_2PbNb_2O_9$ [7-9]. In contrast, a direct, discontinuous, transition from the ferroelectric $A2_1am$ to paraelectric

I4/mmm phase occurs in $Bi_2SrNb_2O_9$, with no evidence for an intermediate non-polar orthorhombic Amam phase [10]. The factors responsible for stabilising the intermediate Amam phase are, at present, unclear. In order to investigate this issue in more detail we have undertaken a high resolution time-of-flight neutron powder diffraction study of the n = 2 Aurivillius phase $Bi_3Ti_{1.5}W_{0.5}O_9$. Although this compound was first reported by Kikuchi in 1977 [12], the structure of this compound had not been determined prior to this study. Our investigation has determined that $Bi_3Ti_{1.5}W_{0.5}O_9$ adopts the polar orthorhombic space group $A2_1am$ at room temperature and transforms directly to the non-polar tetragonal space group I4/mmm, above $T_c = 735\ ^\circ C$.

EXPERIMENTAL

A polycrystalline sample of $Bi_3Ti_{1.5}W_{0.5}O_9$ was prepared by solid state reaction of stoichiometric quantities of Bi_2O_3, TiO_2 and WO_3. An intimate mixture of these oxides was reacted at 800°C for 7 h and then at 1100°C for 24 h, with intermittent grindings. X-ray powder diffraction data were acquired at room temperature using a Stoe STADI/P diffractometer equipped with a Ge (111) curved single crystal monochromator, affording Cu $K\alpha_1$ radiation, and operating in transmission mode. Variable temperature time-of-flight neutron powder diffraction data were acquired on the High Resolution Neutron Powder Diffractometer (HRPD) at the ISIS Facility, UK. Data collection times were approximately nine hours in the ferroelectric phase at room temperature, four hours in the paraelectric phase at 750°C and twenty minutes at intermediate temperatures. The crystal structures of $Bi_3Ti_{1.5}W_{0.5}O_9$ at room temperature and 750°C were determined by Rietveld analysis using the GSAS suite of programs [13]. For the purpose of structure refinement neutron diffraction data from the high resolution back-scattering detectors ($0.6 < d < 4.6$ Å, $\Delta d / d = 1.5 \times 10^{-4}$) and transverse 90° detectors ($0.9 < d < 6.6$ Å, $\Delta d / d = 2 \times 10^{-3}$) were analysed. The temperature dependence of the relative permittivity of $Bi_3Ti_{1.5}W_{0.5}O_9$ was determined by measuring the capacitance of a sintered pellet, of known dimensions, using an LCR meter and a frequency of 100 KHz.

RESULTS

The synthesis of single phase polycrystalline $Bi_3Ti_{1.5}W_{0.5}O_9$ was confirmed by X-ray powder diffraction. Preliminary analysis of X-ray and neutron powder diffraction patterns acquired at room temperature indicated that all reflections could be indexed on the basis of an orthorhombic unit cell in space group $A2_1am$, with $a = 5.402$ Å, $b = 5.372$ Å and $c = 24.94$ Å, in agreement with an earlier study [14]. An initial fit to the neutron diffraction data using a model based on the crystal structure of $Bi_2SrTa_2O_9$ proved satisfactory. The room temperature crystal structure of $Bi_3Ti_{1.5}W_{0.5}O_9$ was determined by refinement of the structural model against the combined X-ray and neutron powder diffraction data sets. The co-ordinates and isotropic thermal parameters of the Ti and W atoms were constrained to adopt the same values in the refinement. The final model provided an excellent simultaneous fit to both the X-ray and neutron diffraction data, see Figure 1. The refined structural parameters are given in Table I and the room temperature crystal structure of $Bi_3Ti_{1.5}W_{0.5}O_9$ is shown in Figure 2.

The principal bond lengths in $Bi_3Ti_{1.5}W_{0.5}O_9$ are given in Table II. The Bi1 cation in the perovskite type layer has a distorted 8-fold co-ordination environment according to the sum of the effective ionic radii, this is presumably due to the steric requirement of the $6s^2$ lone pair electrons. The $(Ti,W)O_6$ octahedra undergo co-operative tilting, due partly to the asymmetric co-ordination of the Bi1 cation and also to bond length mismatch at the interface of the fluorite and perovskite layers. Armstrong and Newnham have proposed a model to quantify the interfacial strain arising from bond length mismatch at the interface between the perovskite and fluorite type layers in Aurivillius phases [15]. These authors suggest the "ideal" *a*-parameter of the fluorite type $Bi_2O_2^{2+}$ layer to be 3.80 Å, tolerance factor calculations for the perovskite type $BiTi_{1.5}W_{0.5}O_7^{2+}$ layer, on the basis of the model proposed by Armstrong and Newnham, afford an ideal *a*-parameter of 3.96 Å. The $(Ti,W)O_6$ octahedra undergo co-operative rotation in order to accommodate the interfacial mismatch between the perovskite and fluorite type layers, whilst maintaining reasonable (Ti,W)-O bond lengths. The Bi2 cation in the fluorite type layer is co-ordinated to four oxygen anions in the form of a distorted square pyramid with two longer Bi2-O3 bonds formed with the apical oxygen anions of the $(Ti,W)O_6$ octahedra. The Bond Valence Sums [16] of the Bi1, Bi2 and Ti / W sites were calculated as 2.8+, 2.8+ and 4.4+, the anticipated mean Bond Valence Sum of the mixed Ti / W site is 4.3+. These Bond Valence Sums indicate that the Bi cations are slightly under-onded in the crystal structure of $Bi_3Ti_{1.5}W_{0.5}O_9$, presumably due to the distortion of the Bi co-ordination environment required to accommodate the $6s^2$ lone pair.

Table I : Refined structural parameters for $Bi_3Ti_{1.5}W_{0.5}O_9$ at room temperature.

Space group:			$A2_1am$		
Unit cell parameters:	a = 5.40181(2) Å		b = 5.37274(2) Å		c = 24.9388(1) Å
Atom	x	y	z	B_{iso} ($Å^2$)	n
Bi1	0*	0.2588(4)	0	2.05(3)	1
Bi2	0.5116(5)	0.7662(2)	0.19803(2)	1.26(2)	1
Ti1	0.5493(7)	0.746(1)	0.41188(8)	0.33(6)	0.75
W1	0.5493(7)	0.746(1)	0.41188(8)	0.33(6)	0.25
O1	0.5815(6)	0.3104(5)	0	0.89(6)	1
O2	0.7794(6)	0.9975(3)	0.25063(8)	0.94(3)	1
O3	0.5596(5)	0.7034(4	0.34134(5)	1.36(4)	1
O4	0.7786(6)	0.9690(4)	0.06999(6)	1.11(4)	1
O5	0.8540(5)	0.9581(4)	0.58552(7)	1.09(5)	1
Powder statistics		R_{wp}	R_p		
Neutron: back-scattering		5.39	5.02		
Neutron: transverse		4.18	3.44		
X-ray		12.40	9.56		
Total		5.89	8.97		$\chi^2 = 3.02$

* Fixed to define origin of polar axis.

Table II : Principal bond lengths for $Bi_3Ti_{1.5}W_{0.5}O_9$ at room temperature.

Fluorite type layer (Å)		**Perovskite type layer (Å)**	
Bi2-O2	2.314(3)	Bi1-O1	2.277(3)
Bi2-O2	2.214(3)	Bi1-O1	2.356(4)
Bi2-O2	2.284(3)	Bi1-O4	2.627(2) x 2
Bi2-O2	2.411(3)	Bi1-O4	2.609(3) x 2
Bi2-O3	2.636(3)	Bi1-O5	2.513(2) x 2
Bi2-O3	2.559(2)	Ti/W-O1	2.231(2)
		Ti/W-O3	1.775(2)
		Ti/W-O4	1.920(5)
		Ti/W-O4	1.990(6)
		Ti/W-O5	1.907(6)
		Ti/W-O5	2.001(4)

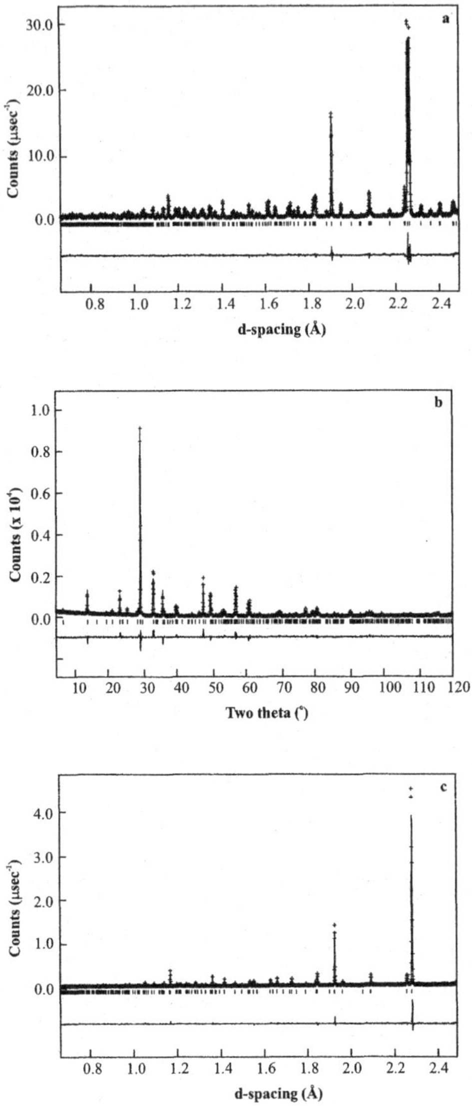

Figure 1: Final profile fit (solid line) to a) neutron and b) X-ray powder diffraction data (solid points) for $Bi_3Ti_{1.5}W_{0.5}O_9$ at room temperature and c) and neutron diffraction data (solid points) at 750°C. The tick marks show the allowed positions of the Bragg reflections, the difference profile is shown below.

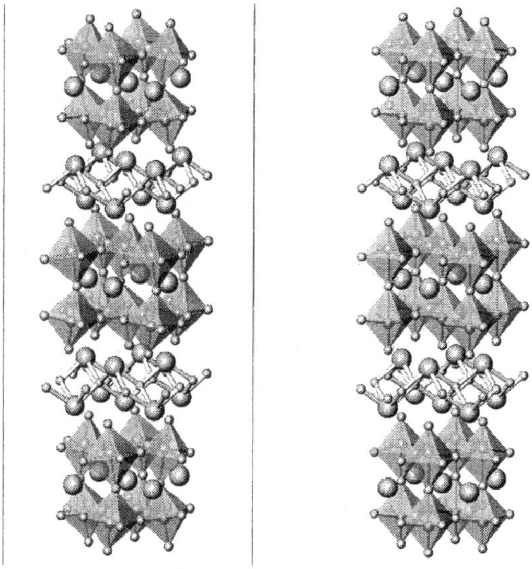

$A2_1am$: 25°C $\quad$ I4/mmm: 750°C

Figure 2: Crystal structures of $Bi_3Ti_{1.5}W_{0.5}O_9$ at 25°C and 750°C, large spheres denote bismuth atoms and small spheres oxygen atoms, note the co-operative tilting of $(Ti,W)O_6$ octahedra in the ferroelectric $A2_1am$ phase.

At room temperature the Ti/W cations are displaced from the centre of the $(Ti,W)O_6$ octahedra toward the O3 anions forming short Ti/W-O3 and longer Ti/W-O1 apical bonds, with four near equivalent equatorial Ti/W-O bonds. This asymmetric co-ordination environment is a common feature of transition metal oxides in which the transition metal possesses a d^0 electron configuration [17] and arises from a second order Jahn Teller effect [18].

The temperature dependence of the crystal structure of $Bi_3Ti_{1.5}W_{0.5}O_9$ was studied by variable temperature high resolution neutron powder diffraction, between room temperature and 750°C. These data were not of sufficient quality to allow the crystal structure of $Bi_3Ti_{1.5}W_{0.5}O_9$ to be refined at each temperature but sufficed to allow the precise measurement of lattice parameters by full pattern fitting using the GSAS suite of programs. The temperature dependence of the unit cell parameters of $Bi_3Ti_{1.5}W_{0.5}O_9$ show that the structure becomes tetragonal above 700°C, within experimental precision, see Figure 3. However, the temperature dependence of the relative permittivity of $Bi_3Ti_{1.5}W_{0.5}O_9$, Figure 4, indicates that ferroelectric – paraelectric phase transition in this material occurs at T_c = 735°C. Close inspection of the neutron powder diffraction patterns collected between 700 – 750°C revealed the presence of supercell reflections associated with space group $A2_1am$ up to 730°C. In contrast, these reflections were absent from the patterns acquired at 740°C and 750°C. This is demonstrated by Figure 5, which shows the persistence of the (4 1 1), (1 4 0), (1 2 16) and (0 3 13) reflections to 730°C. The crystal structure of $Bi_3Ti_{1.5}W_{0.5}O_9$ therefore remains crystallographically orthorhombic, in space group $A2_1am$, to 730°C, but becomes metrically "tetragonal" above 700°C, i.e. $a = b$ within experimental precision. True tetragonal symmetry, space group I4/mmm, is adopted above the ferroelectric –

paraelectric transition temperature, T_c = 735°C. These data indicate that the ferroelectric – paraelectric phase transition in $Bi_3Ti_{1.5}W_{0.5}O_9$ occurs directly from space group $A2_1am$ to I4/mmm at 735°C and is discontinuous in nature. The possibility of an intermediate non-polar orthorhombic Amam phase, characteristic of a continuous transition, is effectively eliminated given the small temperature range over which this phase could be stabilised, approximately 5°C.

The crystal structure of $Bi_3Ti_{1.5}W_{0.5}O_9$ was determined at 750°C in the centrosymmetric tetragonal space group I4/mmm, by refinement against neutron powder diffraction data. The co-ordinates and isotropic thermal parameters of the Ti and W atoms were constrained to adopt the same values in the refinement, as before. The final model provided an excellent fit to the neutron diffraction data as shown in Figure 1. The refined structural parameters are given in Table III and the crystal structure of $Bi_3Ti_{1.5}W_{0.5}O_9$ at 750°C is shown in Figure 2.

Table III : Refined structural parameters for $Bi_3Ti_{1.5}W_{0.5}O_9$ at 750°C.

Space group:	I4/mmm				
Unit cell parameters:	a = 3.85042(2) Å		c = 25.0928(2) Å		
Atom	x	y	z	B_{iso} (Å^2)	n
Bi1	0	0	0	5.61(1)	1
Bi2	0	0	0.19808(5)	3.48(1)	1
Ti1	0	0	0.4077(4)	4.73(4)	0.75
W1	0	0	0.4077(4)	4.73(4)	0.25
O1	0	0	0.5	5.29(1)	1
O2	0	0	0.34188(1)	5.23(1)	1
O3	0	0.5	0.25	2.58(2)	1
O4	0	0.5	0.42292(8)	5.06(1)	1
O5	0	0	0	5.6(1)	1
Powder statistics		R_{wp}	R_p		
Back-scattering		5.47	5.42		
Transverse		4.38	4.30		
Total		4.81	4.58		χ^2 = 1.20

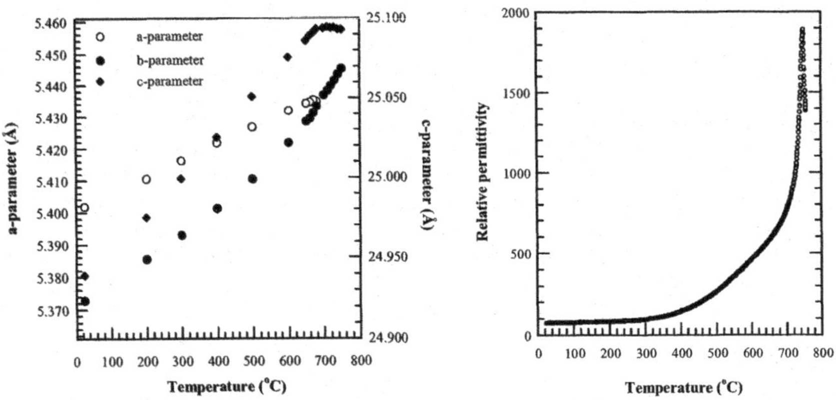

Figure 3: Temperature dependence of unit cell parameters of $Bi_3Ti_{1.5}W_{0.5}O_9$ (above left), data above T_c = 735°C correspond to $a_{tet}\sqrt{2}$.

Figure 4: Temperature dependence of the relative permittivity of $Bi_3Ti_{1.5}W_{0.5}O_9$ (above right).

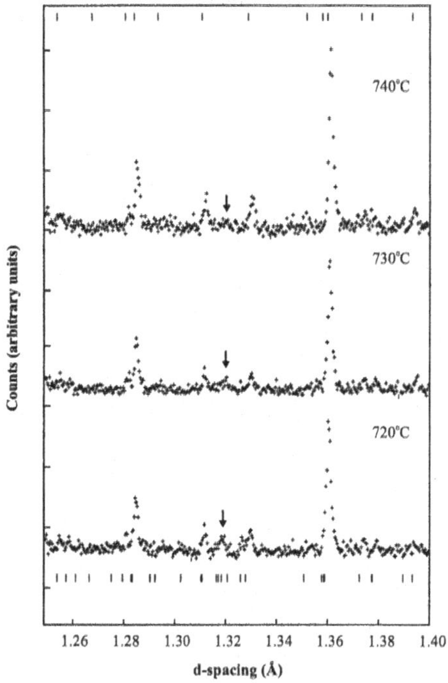

Figure 5: Neutron powder diffraction data for $Bi_3Ti_{1.5}W_{0.5}O_9$, showing the presence of the (4 1 1), (1 4 0), (1 2 16) and (0 3 13) reflections to 730°C and the absence of these reflections at 740°C. The tick marks denote the allowed Bragg reflections in space groups $A2_1am$ (lower set) and I4/mmm at (upper set).

The displacement of the Bi1 cations in the perovskite layer and co-operative rotation of the $(Ti,W)O_6$ octahedra observed in the ferroelectric $A2_1am$ phase of $Bi_3Ti_{1.5}W_{0.5}O_9$ are lost on the transition to the paraelectric I4/mmm phase, as shown in Figure 2. The asymmetric co-ordination environment of the Ti/W cations in the $(Ti,W)O_6$ octahedra is retained in the paraelectric I4/mmm phase, this asymmetric co-ordination environment arises from a second-order Jahn Teller effect as a consequence of the d^0 electron configuration of the Ti and W cations.

DISCUSSION

The currently available data concerning the ferroelectric to paraelectric phase transitions in n = 2 Aurivillius phases are summarised in Table IV. At least two types of behaviour can be distinguished on the basis of these data. The first class of behaviour, demonstrated by $Bi_3Ti_{1.5}W_{0.5}O_9$ and $Bi_2SrNb_2O_9$, involves a direct, discontinuous, phase transition between space groups $A2_1am$ and I4/mmm. In both cases, the materials become metrically "tetragonal" over a narrow temperature range, ~ 40°C, prior to the phase transition, but retain true orthorhombic symmetry. The second class of behaviour demonstrated by $Bi_2SrTa_2O_9$, $Bi_{2.1}Sr_{0.85}Ta_2O_9$ and

$Bi_2PbNb_2O_9$, involves a continuous phase transition between space groups $A2_1am$ and I4/mmm *via* the intermediate space group Amam. The behaviour of $Bi_2PbTa_2O_9$ is similar, although the $A2_1am$ phase becomes metrically "tetragonal" prior to T_c. In all cases the intermediate Amam phase is stable over a considerable temperature range, between 75 – 250°C.

The data in Table IV do not indicate any simple correlation between crystal chemistry and phase transitions in $Bi_2AB_2O_9$ Aurivillius phases. The origin of ferroelectric behaviour in materials of this type was determined by Withers *et al.* to be the displacement of the A cation in the perovskite layer as a consequence of under-bonding [19]. The $A2_1am$ to Amam phase transition involves loss of the A site cation displacement and the tilt mode about the *c*-axis, the remaining tilt mode being retained. The presence of octahedral tilting in perovskite type oxides is symptomatic of bond length mismatch, as outlined above. The presence of octahedral tilting in the Amam phase suggests the presence of residual structural strain, arising from bond length mismatch, in materials adopting this structure above T_c. The stability of the intermediate Amam phase is therefore expected to decrease with increasing T_c, since at higher Curie temperatures thermal expansion will serve to reduce bond length mismatch. This hypothesis is in agreement with the data summarised in Table IV. The observation of a direct $A2_1am$ to I4/mmm transition in the case of $Bi_3Ti_{1.5}W_{0.5}O_9$ may therefore be rationalised in terms of the rather high T_c = 735°C of this material, such that the Amam stability window collapses to zero or is at least very small (certainly $\leq$5°C).

Table IV : Available data for phase transitions in $Bi_2AB_2O_9$ Aurivillius phases, all materials are orthorhombic in space group $A2_1am$ at room temperature.

Phase	$A2_1am$ "$a=b$"	Amam $a \neq b$	Amam "$a=b$"	I4/mmm	PND/ XRD *	Ref.
$Bi_3Ti_{1.5}W_{0.5}O_9$	700°C	—	—	735°C	PND	This study
$Bi_2SrNb_2O_9$	400°C	—	—	440°C	PND	[10]
$Bi_2SrTa_2O_9$	—	300°C		550°C	PND	[9]
$Bi_{2.1}Sr_{0.85}Ta_2O_9$	—	375°C		550°C	PND	[7]
$Bi_2PbNb_2O_9$		575°C	625°C	650°C	SXRD	[8]
$Bi_2PbTa_2O_9$	200°C	—	400°C	500°C	SXRD	[8]

* Denotes the analysis of neutron powder (PND) or synchrotron X-ray (SXRD) powder diffraction data.

CONCLUSIONS

$Bi_3Ti_{1.5}W_{0.5}O_9$ adopts the polar orthorhombic space group $A2_1am$ at temperatures below the ferroelectric Curie temperature, T_c = 735°C. A direct transition to the non-polar tetragonal space group I4/mmm is observed on passing through T_c, with no evidence of an intermediate non-polar Amam orthorhombic phase. Analysis of high resolution neutron powder diffraction data has shown that $Bi_3Ti_{1.5}W_{0.5}O_9$ becomes metrically "tetragonal" above 700°C but remains crystallographically orthorhombic, as demonstrated by the presence of weak supercell reflections associated with space group $A2_1am$. A consideration of the available diffraction data concerning the ferroelectric – paraelectric phase transition in double layered $Bi_2AB_2O_9$ Aurivillius phases leads us to propose that the interplay between A-site cation displacement and bond length mismatch may be critical in controlling the stability of the non-polar Amam orthorhombic phase observed as an intermediate in the ferroelectric – paraelectric phase transition of some $Bi_2AB_2O_9$ phases.

ACKNOWLEDGEMENTS

We thank Dr. I.M. Reaney (The University of Sheffield) for useful discussions and the EPSRC for provision of direct access beam time at the ISIS Facility.

REFERENCES

1. J.F. Scott, "Ferroelectric memories", *Phys. World,* **Feb.** 47 (1995).
2. C. A-Paz de Araujo, J.D. Cuchiaro, L.D. McMillan, M.C. Scott and J.F. Scott, "Fatigue-free ferroelectric capacitors with platinum-electrodes", *Nature*, **374** 627 (1995).
3. B.H. Park, B.S. Kang, S.D. Bu, T.W. Noh, J. Lee, W. Jo, "Lanthanum-substituted bismuth titanate for use in non-volatile memories", *Nature*, **401** 682 (1999).
4. A. Kingon, "Device physics - Memories are made of ...", *Nature*, **401** 659 (1999).
5. B. Aurivillius, "Mixed bismuth oxides with layer lattices. I. Structure type of $CaCb_2Bi_2O_9$", *Ark. Kemi.*, **1** 463 (1949).
6. B. Aurivillius, "Mixed bismuth oxides with layer lattices. II. Structure of $Bi_4Ti_3O_{12}$", *Ark. Kemi.*, **1** 499 (1949).
7. C.H. Hervoches, J.T.S. Irvine, and P. Lightfoot, "Two high-temperature paraelectric phases in $Sr_{0.85}Bi_{2.1}Ta_2O_9$", *Phys. Rev. B.*, **64** 100102 (2001).
8. R. Macquart, B.J. Kennedy, B.A. Hunter and C.J. Howard, "High-temperature structural studies of $PbBi_2M_2O_9$ (M = Nb and Ta)", *J. Phys. Condens. Mat.*, **14** 7955 (2002).
9. R. Macquart, B.J. Kennedy, B.A. Hunter, C.J. Howard and Y. Shimakawa, "Structural phase transitions in the ferroelectric oxide $SrBi_2Ta_2O_9$", *Integr. Ferroelectr.*, **44** 101 (2002).
10. A. Snedden, C.H. Hervoches and P. Lightfoot, "Ferroelectric phase transitions in $SrBi_2Nb_2O_9$ and $Bi_5Ti_3FeO_{15}$: A powder neutron diffraction study", *Phys. Rev. B.*, **67** 092102 (2003).
11. I.M. Reaney, M. Roulin, H.S. Shulman and N. Setter, "*In situ* observations of octahedral tilt transitions in strontium bismuth titanate layered perovskites", *Ferroelectrics*, **165** 295 (1995).
12. T. Kikuchi, "Synthesis of new, layered bismuth titanates, $Bi_7Ti_4NbO_{12}$ and $Bi_6Ti_3WO_{18}$", *J. Less Common Met.*, **52** 163 (1977).
13. A.C. Larson and R.B. Von Dreele, "GSAS – General Structural Analysis System", Report LA-UR-86-748, Los Alamos National Laboratory, Los Alamos, NM 87545, 1990.
14. S. Luo, Y. Noguchi, M. Miyayama and T. Kudo, "Rietveld analysis and dielectric properties of $Bi_2WO_6 - Bi_4Ti_3O_{12}$ ferroelectric system", *Mat. Res. Bull.*, **36** 531 (2001).
15. R.A. Armstrong and R.E. Newnham, "Bismuth titanate solid solutions", *Mat. Res. Bull.*, **7** 1023 (1972).
16. I.D. Brown and S. Altermatt, "Bond-valence parameters obtained from a systematic analysis of the Inorganic Crystal Structure Database", *Acta Cryst.*, **B41** 244 (1985).
17. N.S.P. Bhuvanesh and J. Gopalakrishnan, "Solid state chemsitry of early transition-metal oxides containing d^0 and d^1 cations", *J. Mat. Chem.*, **7** 2297 (1997).
18. R.A. Wheeler, M.H. Whangbo, T. Hughbanks, R. Hoffmann, J.K. Burdett and T.A. Albright, "Symmetric vs. asymmetric linear M-X-M linkages in molecules, polymers and extended networks", *J. Am. Chem. Soc.*, **108** 2222 (1986).
19. R.L. Withers, J.G. Thompson and A.D. Rae, "The crystal chemistry underlying ferroelectricity in $Bi_4Ti_3O_{12}$, Bi_3TiNbO_9 and Bi_2WO_6", *J. Solid State Chem.*, **94** 404 (1991).

ELECTRIC FIELD-INDUCED INTERSECTIONS OF 90° DOMAIN WALLS IN TETRAGONAL FERROELECTRIC CRYSTALS

Xiaoli Tan
Dept. Mater. Sci. & Eng.
Iowa State University
Ames, IA 50011

Jian-Ku Shang
Dept. Mater. Sci. & Eng.
Univ. Illinois at Urbana-Champaign
Urbana, IL 61801

ABSTRACT

Intersections of the permissible 90° domain walls in tetragonal crystals were analyzed and the fifteen combinations were classified into two major groups. Group A is characterized by a 90° intersecting angle and a <001> intersecting line while group B is characterized by a 60° intersecting angle and a <111> intersecting line. When viewed along the $<00\bar{1}>$ direction of thin crystals, the projected images of these intersections appear in four configurations. Electrical charges due to the termination of domains and elastic distortion due to the tetragonality of the lattice structure are introduced at the intersections. Three out of the four configurations of intersections were experimentally identified. The one absent was interpreted as unfavorable under in-plane electric fields. The intersection was observed to lead to the formation of partial dislocations and microcracks. It is proposed that the atomic shifts during 90° domain switching initiate the dislocation formation and the stored elastic energy at intersections is responsible for the microcrack formation.

INTRODUCTION

Ferroelectric domains are the volumetric units with homogeneous electrical polarization. They respond to external electrical stimulus by the expansion of domains with favored polarization orientations. Domain walls and domain configurations have been observed to significantly impact materials' dielectric and piezoelectric properties. It was found in barium titanate ceramics that domain wall motion contributed to the dielectric constant on the order of 40%.[1] In $Pb(Mg_{1/3}Nb_{2/3})O_3$–$PbTiO_3$ (PMN–PT) and $Pb(Zn_{1/3}Nb_{2/3})O_3$–$PbTiO_3$ (PZN–PT) single crystals, ultrahigh piezoelectric properties were observed only in domain-engineered crystals in multidomain state.[2-4]

In tetragonal ferroelectric crystals, the polarization vector is along one of their prototype cubic <001> directions, and the rotation of the polarization vector from one domain to the next can only be 90° or 180°. Therefore, 90° domains and 180° domains exist in tetragonal crystals. In literature, ferroelectric domains with in-plane polarization vectors in tetragonal crystals are often referred to as *a*-domains and the ones with out-of-plane vectors are termed as *c*-domains.[5] The permissible uncharged 90° walls are along the {101} plane in tetragonal crystals. In a {010} thin foil, the 90° wall separating two *a*-domains is perpendicular to the surface plane while the wall separating an *a*-domain from a *c*-domain is 45° inclined to the surface plane. In TEM, therefore, the *a*/*a*-domain wall appears as a thin line while the inclined *a*/*c* domain wall appears with fringe contrast.[5]

Application of electric fields can cause a transformation of the domain structure, leading to the formation of complex patterns. The complex domain structures often include intersecting domain walls.[5-11] As a result of the intersection, electric as well as mechanical complications may arise in the ferroelectric response of a multidomained structure. To the electric behavior, domain intersections carry excess electrical charges and thus provide a convenient path for charge injection which in turn may cause electric fatigue.[9] To the mechanical response, the intersection produces local lattice distortion, which may result in changes in domain wall density[7] or local fracture.[12,13]

The intersections of 90° domain walls have been observed previously in both single crystals[8-13] and polycrystalline ceramics.[5-7] However, in-depth analysis and detailed examinations of the intersections are yet to be performed. We have reported experimental observations on intersections of (i) orthogonal *a*-domains in the *a*-domain matrix;[14] (ii) *c*-domains with *a*-domains in the *a*-domain matrix;[15] and (iii) orthogonal *a*-domains in the *c*-domain matrix.[13] In this article, we summarize the intersections of 90° domain walls in tetragonal ferroelectric crystals and discuss the consequences of these intersections.

EXPERIMENTAL PROCEDURE

Thin {010} slices of tetragonal PMN–PT crystal were obtained from H.C. Materials (Urbana, Illinois). Three sets of specimens were prepared for the examination of 90° domain wall intersections.

For the study of the intersections between orthogonal *a*-domain strips in the *a*-domain matrix, specimens with dimensions of $6.0 \times 3.0 \times 0.8$ mm^3 with surfaces parallel to the pseudo-cubic {010} planes were cut from a thin slice. The two major surfaces (6.0×3.0 mm^2) were polished with diamond paste. In order to remove the residual stress generated during sample preparation, the polished specimens were annealed at 300°C for 3 hours. The specimens were then electroded by evaporating Au to the two side surfaces of 6.0×0.8 mm^2. The Au film electrodes were then connected to a high voltage power supply. Cyclic sinusoidal electric fields of 30 Hz were applied across the specimen along the <100> direction to trigger domain switching. Morphology changes of 90° domains were monitored by optical microscopy under cross-polarized light and images were recorded regularly with the attached charge-coupled-device camera.

The second set specimens were prepared as disks with diameter of 3 mm. These {010} thin disks were ground and polished to a thickness of ~120 µm. The central area of the disks was further thinned to about 15 µm by mechanical dimpling. The dimpled specimens were then annealed at 250°C for 1 hour to minimize residual stresses. Two half-circle shaped Au films were evaporated on the flat polished surfaces of the specimens with a gap of 500 µm along the <001> direction. Bipolar cyclic electric fields were applied to the electroded specimens at an amplitude of ±6.5 kV/cm for 5×10^6 cycles. The electrically cycled specimens were then thinned with argon ion mill at 5 kV and 12° incidence angle until perforation. Domain intersections were examined on a Phillips CM-12 TEM operated at 120kV.

The third set was also prepared as 3-mm-diameter disks. However, they were cut from a thin {010} slice that had been poled along the thickness direction. The sample preparation steps followed the same procedure as that for the second set. X-ray diffraction confirmed a strong (002) peak, indicating a major volume fraction of *c*-domains. After the application of bipolar cyclic fields, optical examination as well as TEM examination was conducted.

COMBINATIONS OF INTERSECTING 90° DOMAIN WALLS

Ferroelectric domains are bounded by domain walls. Therefore, domain intersections in this article refer to two interrelated aspects: the intersecting domain walls and the contact of domain strips. When 90° domains in tetragonal crystals are concerned, the intersecting domain walls are of {101} type crystallographic planes.

There are six equivalent {101} planes in a pseudocubic crystal. Combinations of two {101} planes out of six result in fifteen configurations. These fifteen combinations are listed in Table I, where they are sorted into 4 subgroups according to their appearance in thin foil crystals, as schematically sketched in the last column. This column shows the projected image when viewed along the $\langle 00\bar{1}\rangle$ direction, where the lines with shadow indicate an inclined wall. Symbols ↓, ↑, →, ⊙, and ⊗ represent the polarization vectors. The assignment of the polarization vector to each domain in Table I is based on an assumption that the two intersecting domains have the common matrix domain, such as the *a*-domains denoted by ↑ in the subgroups A_1, A_2 and B_1, and the *c*-domains denoted by ⊙ in the subgroup B_2.

Table I. Possible 90° domain intersections in tetragonal crystals. All the planes and directions are indexed as in cubic structure.

Subgroup	Intersecting domains	Intersecting angle	Intersecting line	$\Sigma\eta_l$	Projected image along [001]
A_1	$(\bar{1}10)(110)$	90°	[001]	$[0\bar{2}0]$	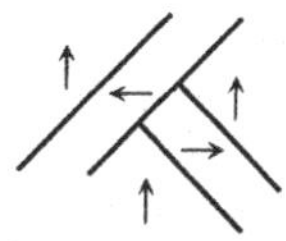
A_2	$(\bar{1}01)(101)$ $(0\bar{1}1)(011)$	90°	$\langle 100\rangle$	$\langle 00\bar{2}\rangle$	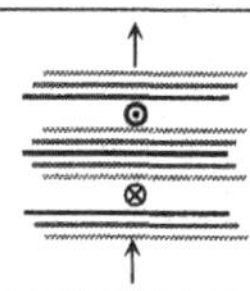
B_1	$(011)(\bar{1}10)$ $(0\bar{1}1)(\bar{1}10)$ $(\bar{1}01)(\bar{1}10)$ $(101)(\bar{1}10)$ $(011)(110)$ $(0\bar{1}1)(110)$ $(\bar{1}01)(110)$ $(101)(110)$	60°	$\langle 1\bar{1}1\rangle$	$\langle 1\bar{2}1\rangle$	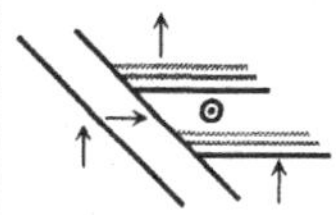
B_2	$(\bar{1}01)(0\bar{1}1)$ $(101)(0\bar{1}1)$ $(\bar{1}01)(011)$ $(101)(011)$	60°	$\langle 1\bar{1}1\rangle$	$\langle 1\bar{2}1\rangle$	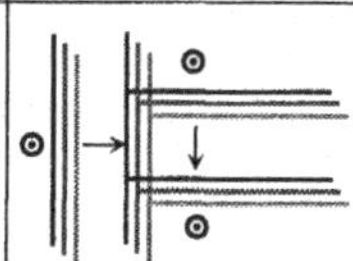

The 90° domain walls in tetragonal ferroelectric crystals bear similarities to mechanical twinning planes when atomic arrangements are concerned, as suggested in our previous study.[8] The domain wall motion involves a shear displacement along the $\langle 10\bar{1}\rangle$ direction of atoms. In analogy with mechanical twinning notation, this shear is denoted as η_I. For atoms that sit on the intersecting lines, they should experience both shears simultaneously. The magnitude of the sum of these two shears is listed in the $\Sigma\eta_I$ column in Table I. The shear along different directions at the intersecting line results in local elastic distortion energy, which is the mechanical consequence of domain intersections.

As shown in Table I, subgroup A_1 represents the intersections of orthogonal *a*-domain strips in the *a*-domain matrix. Subgroup A_2, which has two variants, includes configurations of the intersections between *c*-domain strips with anti-parallel polarizations in the *a*-domain matrix. In the projected image, the traces of the inclined domain walls appear to be parallel to each other. However, subgroups A_1 and A_2 are geometrically identical and they form the major group A. Intersections of domain walls in group A have a 90° intersecting angle and a <001> intersecting line. The superimposed shear is along the <020> direction. Subgroup B_1, with eight combinations, represents the intersections of *c*-domain strips with *a*-domain strips in the *a*-domain matrix. In the projected image, the angle between the domain wall traces appears to be 45°. The last subgroup, B_2, which has four combinations, includes intersections between orthogonal *a*-domain strips in the *c*-domain matrix. The 60° intersecting angle in space appears to be 90° in the projected images. Subgroups B_1 and B_2 form major group B since they are geometrically identical in three dimension space. Group B is characterized by 60° intersecting angles, <111> intersecting lines and <121>-direction shear.

The permissible uncharged 90° domain wall separates two adjacent domains with polarization vectors in the "head-to-tail" configuration. However, this configuration is not preserved across the interface between the two intersecting domains housed in the same matrix domain. For example, the diagram for the subgroup A_1 shown in Table I reveals a "head-to-head" configuration across the wall segment between the two intersecting domain strips. Therefore, the domain intersection carries electrical charges. This is the electrical consequence of domain intersections and will be further illustrated later.

FORMATION PROCESS

Mechanical distortion and electrical charges are present at domain intersections. Such energetic cost hinders the formation of intersections. Therefore, non-parallel strips of 90° domains stay apart in the specimens initially, as observed experimentally. Figure 1 shows the initial domain configurations in one of the first set specimens. In these specimens, one set of parallel 90° domain strips was found to occupy a large portion of the sample surface while an orthogonal set of strips was only found in limited areas. This orthogonal set can be seen in the lower left corner of Fig. 1(a). The optical image shows that these two orthogonal sets of domain strips stay apart. TEM

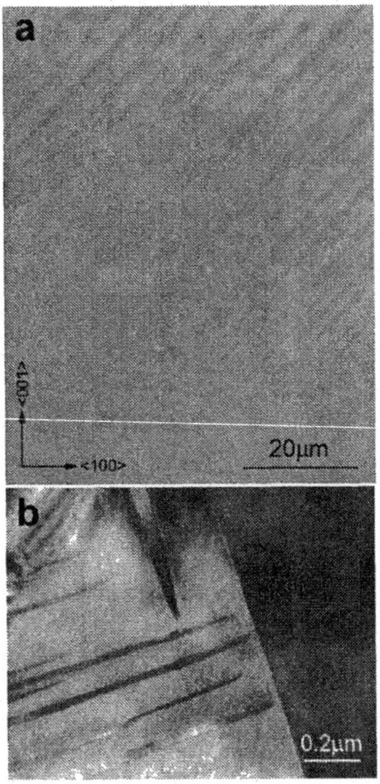

Fig. 1 Initial domain configurations. (a) Optical image of the first set specimens where orthogonal *a*-domains are housed in the *a*-domain matrix. (b) TEM image shows the pointed tips of the orthogonal *a*-domains.

examination of the same specimen clearly indicated that the intersections of 90° domain walls were avoided by forming pointed tips of the *a*-domains, as shown in Fig. 1(b).

However, external stimuli may overcome the energy barrier and force the domains to intersect. Figure 2 shows an example of electric field-induced interpenetrated domain strips, where all the domains, including the matrix domain, are *a*-domains. Therefore, these intersections belong to the subgroup A_1 in Table I. The strips running from upper right to lower left initially occupied the major fraction of the surface. The orthogonal set of strips grew under the applied electric field and was replacing the initial strips. Domain intersections and even interpenetration are generated under external electrical cycling.[14]

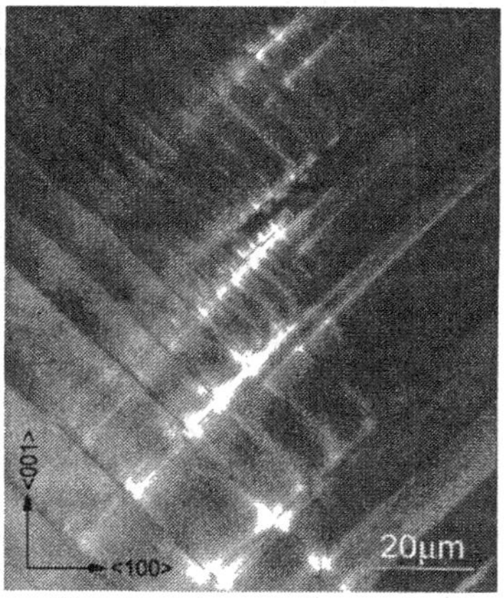

Fig. 2 Interpenetrating *a*-domains intersections. Image was taken after the application of bipolar electric fields for 36,000 cycles at ±4.0 kV/cm and 36,000 cycles at ±5.0 kV/cm.

The formation of domain intersections requires domain expansion. For perovskite ferroelectric crystals, both mechanical and electric stimuli are capable of facilitating domain expansion. The tetragonality of the crystal lattice makes the ferroelectric 90° domains responsive to mechanical stresses and the spontaneous polarization makes the domains responsive to external electric fields. Therefore, both mechanical stresses and electric fields are capable of forming domain intersections.

In our present experimental setup, we applied in-plane electric fields to generate the domain intersections. Therefore, we discuss the formation of domain intersections under in-plane electrical stimuli only. When external electric fields are applied, domains with polarization vectors pointing to the directions close to the field direction will grow. The growth of non-parallel domain strips at the expenses of the matrix domain will lead to the formation of domain intersections. The expansion of the matrix domain will make the non-parallel domain strips move further apart, thus, generates no intersections.

When the external stimulus for the formation of domain intersections is taken into consideration, it will be clear that configuration A_1 is no longer equivalent to A_2, and configuration B_1 is no longer equivalent to B_2 either. The results are summarized in Table II. The first row of Table II schematically shows the initial configuration and the rest two rows illustrate the resulted domain intersections when an electric field is applied. The applied in-plane fields are along the pseudocubic <100> direction, indicated by the long arrows in the table.

We consider subgroup A_1 first. When an electric field is applied along the direction indicated by the arrow in the second row of the table, the domain with pointed tip will grow because its polarization vector is parallel to the field direction. The growth of this domain involves the expansion of its length as well as its width.[9] The matrix domain will be converted to this expanding domain by 90° polarization switching. The domain strip orthogonal to this expanding domain will presumably keep unchanged based on the following arguments. First of all, a 90° switching of the polarization to the matrix domain will not likely to occur since such process would not align the polarization vector to the external field. A 180° polarization switching of this domain would align polarization to the external field, however, it requires a nucleation stage. In addition, even the energy barrier for the nucleation stage is overcome and a 180° switching occurs, the domain wall separating this switched domain from the matrix domain would carry electrical charges since the neutral "head-to-tail" configuration no longer exists. As the expanding domain approaches the impinging orthogonal domain strip, domain intersection ensues. When an electric field is applied along the opposite direction, the orthogonal domain strip will expand at the expense of the matrix domain and the domain with pointed tip will stay unchanged. Similarly, this would lead to the formation of an intersection.

Table II. Domain intersections developed under in-plane electric fields.

	A_1	B_1	B_2
Initial	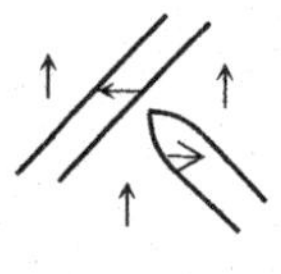	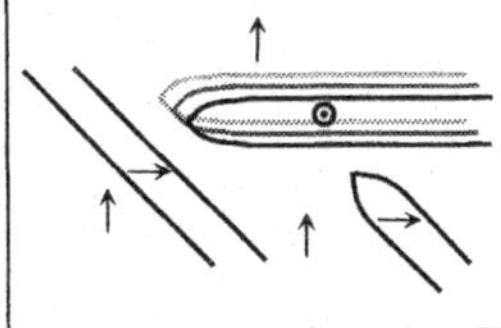	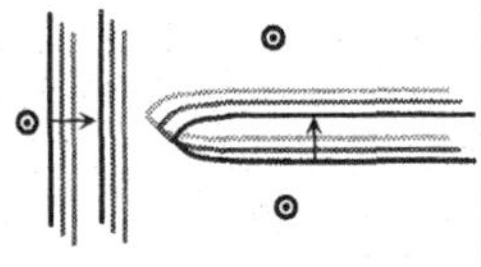
Field on	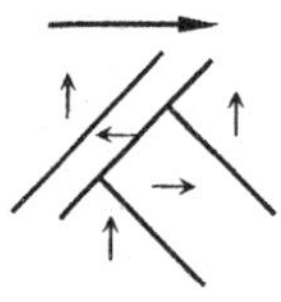	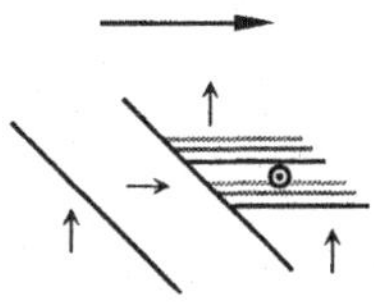	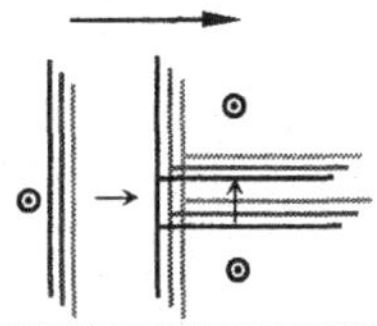
	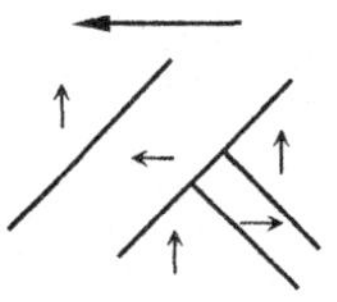	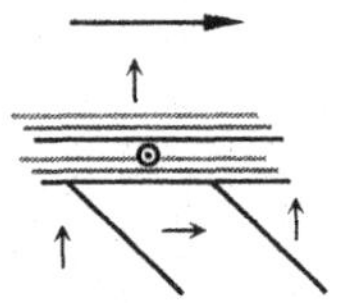	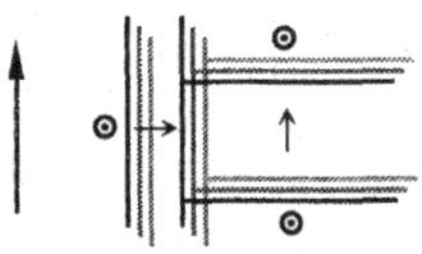

The situation for subgroup A_2 is slightly different. As indicated in Table I, both of the intersecting domains have out-of-plane polarizations. By no means an in-plane electric field would force these two domains grow. The in-plane electric field will only transform these *c*-domains to the matrix *a*-domain and move the domain walls further apart. Therefore, no domain intersections will be induced by electric fields in this situation. The analysis is consistent with our experiments where we have never observed subgroup A_2 intersections in any specimens examined.

For the B_1 configuration, the application of a field along the horizontal direction would lead to the growth of the *a*-domain strips. The *c*-domain would stay unchanged because it is equally unfavored as the matrix domain. When the *a*-domains approach the *c*-domain, domain intersections would be formed. At the intersection, a *c*-domain can terminate at an expanding *a*-domain (illustrated in the second row) or an expanding *a*-domain can terminate at a *c*-domain (illustrated in the third row). For the B_2 configuration, the application of a field along the horizontal direction would lead to the growth of the vertical *a*-domain strip. Similarly, the application of a field along the vertical direction would lead to the growth of the horizontal *a*-domain strip. In either case, domain intersections will be formed.

THE INTERSECTIONS

When the intersecting domains (not the matrix domain) are in contact at the intersection, electrical charges will arise on the segment of the domain wall between intersecting lines.[16] This electrical charge is due to the new polarization vector configuration at the intersection, which will be illustrated in detail in this section. In addition, experimental verification of A_1, B_1, and B_2 intersections will be presented.

3-D Illustration of the Intersection

Analysis in the previous section shows that in-plane electric fields can trigger domain expansion which will in turn lead to the formation of intersections of A_1, B_1 and B_2. In order to illustrate the charged domain wall segment at the intersection, we schematically show the volume in the vicinity of the intersecting lines in three dimensions. Figure 3 illustrates the intersection of subgroup A_1, where two orthogonal *a*-domain strips intersect in the *a*-domain matrix. The line O_1O_2 is the intersecting line of the domain walls. The prism $A_1B_1O_1$–$A_2B_2O_2$ stands for the matrix *a*-domain D_m, the prism $A_1C_1D_1$–$A_2C_2D_2$ depicts the impinging *a*-domain D_i and the prism $B_1C_1O_1$–$B_2C_2O_2$ represents the growing *a*-domain D_a. The rectangular plane $O_1C_1C_2O_2$ is the domain wall segment between D_a and D_i. It is apparent that this segment is a 180° domain wall since D_a and D_i have antiparallel polarization vectors. However, this segment is tilted 45° away from the neutral position and thus carries electrical charges. The charge density on this wall segment is $\sqrt{2}P$, where P is the polarization of the crystal.[16] The electrically neutral domain wall between the two intersecting *a*-domains D_a and D_i is a plane parallel to $A_1B_1B_2A_2$. This plane shares the zone-axis O_1O_2 with the other two permissible uncharged walls ($A_1O_1O_2A_2$ and $O_1B_1B_2O_2$). Therefore, it is possible to eliminate the electrical charges at this intersection by forming a step on the domain wall.[16]

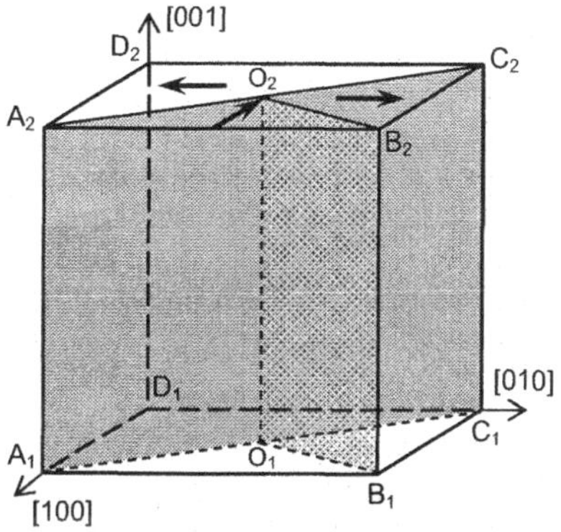

Fig. 3 3-D illustration of the intersection of orthogonal *a*-domains in the *a*-domain matrix.

Figure 4 illustrates the intersecting domain walls in three dimensions for the subgroup B_1. In this diagram, the tetrahedron $B_2B_1C_1D_1$ is the matrix *a*-domain, the prism $A_1B_1D_1$–$A_2B_2D_2$ stands for the growing *a*-domain D_a, while the pyramid B_2–$D_1C_1C_2D_2$ is the impinging *c*-domain D_i. Line D_1B_2 is the intersecting line of the domain walls. The triangular plane $D_1B_2D_2$ is the interface between the D_a and D_i with polarization vectors 90° apart. The polarization vector of D_i is parallel to this segment and the vector of D_a is 45° to it. Therefore, the charge density on this segment is

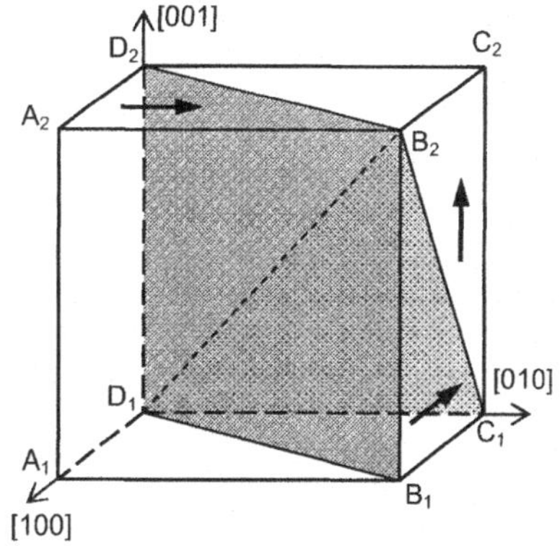

Fig. 4 The intersection of a *c*-domain with an *a*-domain in the *a*-domain matrix.

$\frac{\sqrt{2}}{2}P$. The neutral position of the domain wall between the two intersecting domains would be parallel to the plane $B_1C_1D_2A_2$. However, this plane does not share the zone-axis D_1B_2 with the other two permissible uncharged walls (triangles $D_1B_1B_2$ and $D_1C_1B_2$). Therefore, the electric charges cannot be eliminated by forming a step.

The three dimensional illustration of the subgroup B_2 intersection is shown in Fig. 5. The tetrahedron $B_1C_1B_2A_2$ is the matrix *c*-domain, the prism $A_1B_1A_2$–$D_1C_1D_2$ stands for the growing *a*-domain D_a, while the pyramid C_1–$A_2B_2C_2D_2$ is the impinging *a*-domain D_i. Line C_1A_2 is the intersecting line. The triangular plane $C_1A_2D_2$ is the interface between the D_a and D_i. Similar to the situation in B_1 intersection, this segment carries electrical charges with a density of $\frac{\sqrt{2}}{2}P$. The neutral position of the domain wall between the two intersecting *a*-domains would be parallel to the plane $D_1B_1B_2D_2$. This plane does not share the zone-axis C_1A_2 with the other two permissible uncharged walls (triangles $B_1C_1A_2$ and $B_2C_1A_2$). Therefore, the step mechanism for subgroup A_1 to eliminate charges is not applicable to B_2 configuration either.

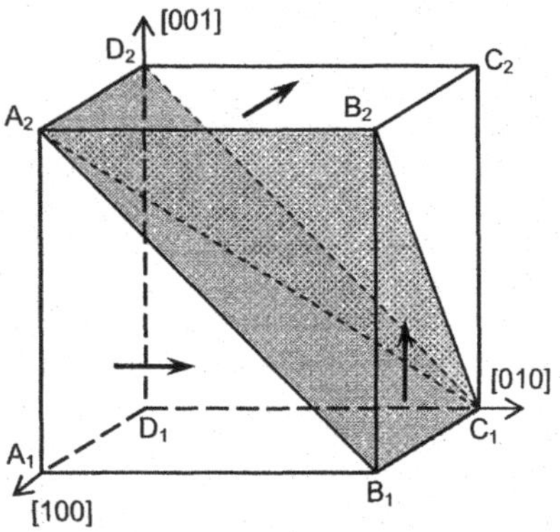

Fig. 5 The intersection of orthogonal *a*-domains in the *c*-domain matrix.

Experimental Confirmation

Our experimental observations reveal the presence of domain intersections of subgroup A_1, B_1, and B_2. Figure 6 shows the intersections of orthogonal strips of *a*-domains in the *a*-domain matrix (subgroup A_1). It is evident from this optical image that the intersecting *a*-domains are in full contact. Figure 7 shows a TEM image of intersections of subgroup B_1, where the growing *a*-domain is contact with four *c*-domains in the *a*-domain matrix. Figure 8 shows the intersections between orthogonal *a*-domains in the *c*-domain matrix, which belong to subgroup B_2.

The net electrical charges will cause extra electrostatic energy in the vicinity of the intersections. The raised energy is a function of the density, distribution, and amount of the charge.[17] In addition, elastic energy will be introduced at the intersections due to the tetragonality of the lattice. For subgroup A_1 intersections, the elastic distortion involves the lattice rotation, while for subgroups B_1 and B_2 it is caused by matching the *a*-axis with the *c*-axis of the tetragonal lattice. The consequences of the raised local energy are the formation of microcracks (Fig. 9) and partial dislocations (Fig. 10) in the vicinity of domain intersections.[13,15]

Fig. 6 Optical image of the intersections of orthogonal *a*-domains in the *a*-domain matrix.

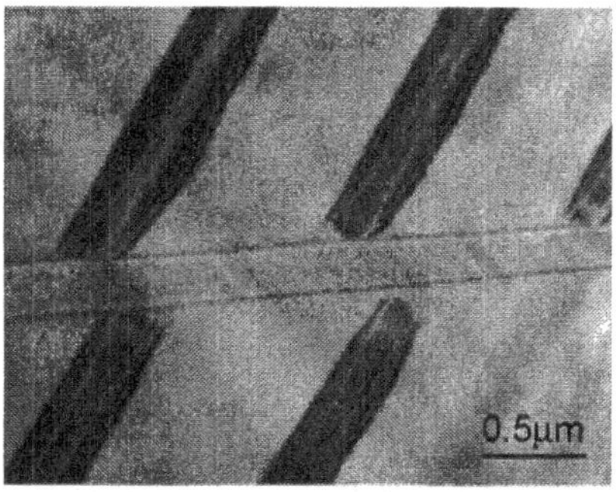

Fig. 7 TEM image of the intersections of *c*-domains with an *a*-domain in the *a*-domain matrix.

Fig. 8 TEM image of the intersections of orthogonal *a*-domains in the *c*-domain matrix.

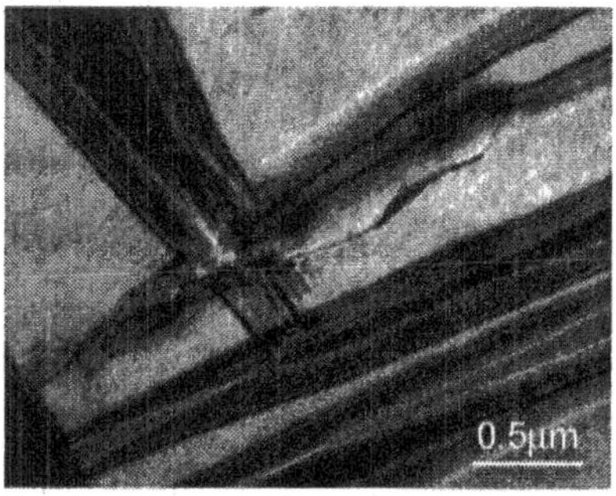

Fig. 9 A microcrack is revealed by TEM at the intersection of two *a*-domains in the *c*-domain matrix.

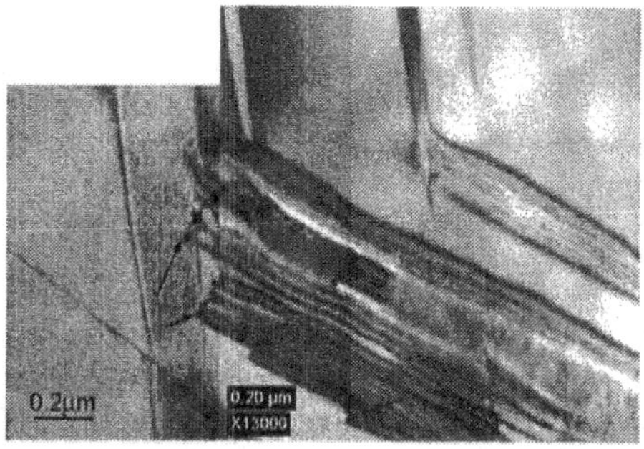

Fig. 10 TEM observations show that partial dislocations were formed at intersections of c-domains with a-domains in the a-domain matrix.

CONCLUSIONS

Four configurations (subgroups A_1, A_2, B_1, and B_2) of intersecting 90° domain walls are possible in <001>-oriented tetragonal ferroelectric crystals. Subgroup A_1 includes intersections resulted from orthogonal *a*-domain strips in the *a*-domain matrix; A_2 includes intersections resulted from parallel *c*-domain strips in the *a*-domain matrix; B_1 refers to intersections resulted from *a*-domains and *c*-domains in the *a*-domain matrix and B_2 denotes intersections resulted from orthogonal *a*-domain strips in the *c*-domain matrix. When an in-plane electrical field is applied, three intersection configurations (A_1, B_1 and B_2) can be generated in a <001>-oriented crystal. Electrical charges are present at intersections due to the termination of ferroelectric domains, while elastic energy is produced at intersections due to the constrained lattice distortion. Experiments have confirmed the formation of microcracks at intersections. In addition, a dislocation mechanism is validated for relieving the stored elastic energy at intersections.

ACKNOWLEDGMENTS

The authors are grateful to Dr. Han for providing single crystals. The experimental work was carried out at the Center for Microanalysis of Materials, University of Illinois, which is supported by the U.S. Department of Energy under grant DEFG02-96-ER45439.

REFERENCES

[1]G. Arlt and H. Peusens, "Dielectric constant of coarse grained $BaTiO_3$ ceramics," *Ferroelectrics*, **48** 213-24 (1983).

[2]S. Park and T. R. Shrout, "Ultrahigh strain and piezoelectric behavior in relaxlor based ferroelectric single crystals," *Journal of Applied Physics*, **82** 1804-11 (1997).

[3]S. Wada, S. Park, L.E. Cross and T.R. Shrout, "Domain configuration and ferroelectric related properties of relaxor based single crystals," *Journal of the Korean Physical Society*, **32** 1290-93 (1998).

[4]D. Viehland, A. Amin and J.F. Li, "Piezoelectric instability in <011>-oriented $Pb(B'_{1/3}B''_{2/3})O_3$–$PbTiO_3$ crystals," *Applied Physics Letters*, **79** 1006-08 (2001).

[5]Y.H. Hu, H. M. Chan, X. W. Zhang and M. P. Harmer, "Scanning electron microscopy and transmission electron microscopy study of ferroelectric domains in doped $BaTiO_3$," *Journal of the American Ceramic Society*, **69** 594-602 (1986).

[6]Y.F. Tsun and C.C. Chou, "Microstructural evolution under polarization switching of ferroelectric tin-modified lead zirconate titanate ceramics," *Japanese Journal of Applied Physics*, **38** 3585-88 (1999).

[7]G. Arlt, and P. Sasko, "Domain configuration and equilibrium size of domains in $BaTiO_3$ ceramics," *Journal of Applied Physics*, **51** 4956-60 (1980).

[8]J.K. Shang and X. Tan, "Indentation-induced domain switching in $Pb(Mg_{1/3}Nb_{2/3})O_3$ –$PbTiO_3$ crystal," *Acta materialia*, **49** 2993-99 (2001).

[9]A. Krishnan, M.E. Bisher and M.M. J. Treacy, "*In-situ* TEM study of domain propagation in ferroelectric barium titanate, and its role in fatigue," *MRS Symp. Proc.*, **541**, 475-80 (1999).

[10]B.M. Park and S.J. Chung, "Optical, electron microscopic, and x-ray topographic studies of ferroic domains in barium titanate crystals grown from high-temperature solution," *Journal of the American Ceramic Society*, **77** 3193-201 (1994).

[11]J. Yin and W. Cao, "Domain configurations in domain engineered $0.955Pb(Zn_{1/3}Nb_{2/3})O_3$-$0.045PbTiO_3$ single crystals," *Journal of Applied Physics*, **87** 7438-41 (2000).

[12]C.T.A. Suchicital, D.A. Payne and A. G. de l'Eprevier, "Crystal growth, phase transformation studies, and micro-cracking in single crystal lead titanate," *ISAF'86*, IEEE, 1986, pp.465-68.

[13]X. Tan and J.K. Shang, "Intersection of *a*-domains in the *c*-domain matrix driven by electric field in tetragonal ferroelectric crystal," *Journal of Applied Physics*, in press (2004).

[14]X. Tan and J.K. Shang, "Field-induced domain interpenetration in tetragonal ferroelectric crystal," *Journal of Applied Physics*, **95** 635-39 (2004).

[15]X. Tan and J.K. Shang, "Partial dislocations at domain intersections in a tetragonal ferroelectric crystal," *Journal of Physics: Condensed Matter*, **16** 1455-66 (2004).

[16]A. Krishnan, M.M.J. Treacy, M.E. Bisher, P. Chandra and P.B. Littlewood, "Maxwell charge on domain walls," *Fundamental Physics of Ferroelectrics 2000*, R.E. Cohen, ed., American Institute of Physics, **535** 191-200 (2000).

[17]B.V. Selyuk, "Charged domain boundaries in ferroelectric crystals," *Ferroelectrics*, **6** 37-40 (1973).

STRUCTURAL TRANSFORMATIONS IN BISMUTH TITANATES

Sudhanshu Mallick, Keith. J. Bowman, Elliott. B. Slamovich, Alexander H. King and Jacob. L. Jones*

School of Materials Engineering
Purdue University
501 Northwestern Avenue
West Lafayette, IN 47907

ABSTRACT

A transformation from the $Bi_4Ti_3O_{12}$ structure to the $Na_{0.5}Bi_{4.5}Ti_4O_{15}$ structure was observed on heating sodium doped bismuth titanate in air. The transformation has been observed to proceed through the mixed-layer structure, $Na_{0.5}Bi_{8.5}Ti_7O_{27}$. High resolution TEM was done to study the structure of bismuth titanate and mechanisms are proposed for the structural transformation.

INTRODUCTION

Bismuth titanate has a layer type structure consisting of perovskite-like blocks interleaved with layers of $(Bi_2O_2)^{2+}$. It can be represented by a general formula $(Bi_2O_2)^{2+}$ $(A_{m-1}B_mO_{3m+1})^{2-}$ where A is a combination of ions coordinated to 12 ions (e.g., Bi, Na, Sr) and B is a combination of ions suitable for 6-fold coordination (e.g. Ti, Nb) [1]. Based on the number of the perovskite-like blocks (quantity m-1) stoichiometries such as $Bi_4Ti_3O_{12}$, $Na_{0.5}Bi_{4.5}Ti_4O_{15}$, etc. are possible [2] (Figure 1). Stoichiometries other than $Bi_4Ti_3O_{12}$ are possible only when bismuth titanate is doped to maintain charge neutrality. Another mixed-layer structure consisting of alternating 2 and 3 perovskite like blocks interleaved with $(Bi_2O_2)^{2+}$ layers has been observed by Kikuchi *et al.* [1,3,4].

Though no reports have been found for bismuth titanate indicating the existence of multiple structures within the same grain, multiple polytypes have been observed within a single grain in the $MgF_2 - Nb_2O_5$ system which has a crystal structure very similar to the Aurivillius phases [5, 6, 7]. Hutchison *et al.* report that the number of perovskite-like blocks between the $Bi_2O_2^{2+}$ layers is either 3 or 4 varying arbitrarily across the sample [8]. Boullay *et al.* have also observed long range order in the intergrowth structure in the $Bi_3TiNbO_9 - Bi_4Ti_3O_{12}$ system [9]. Though they develop the structural relationship between the various Aurivillius phases based on crystal symmetry, the mechanism of formation of the mixed-layer phases has still not been established. The present work demonstrates the structural transformations that occur in the $Bi_4Ti_3O_{12} - Na_{0.5}Bi_{4.5}Ti_4O_{15}$ system and proposes mechanisms which might be responsible for this transformation.

* Current address: Materials Science and Engineering, The University of New South Wales, Sydney 2052, Australia.

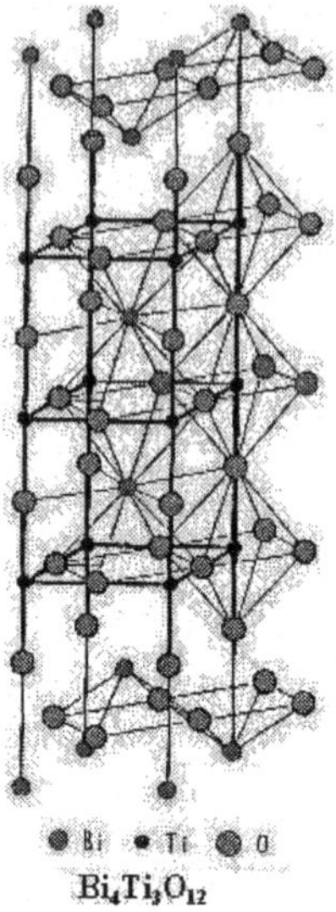

Figure 1: Half unit cell of $Bi_4Ti_3O_{12}$

EXPERIMENTAL PROCEDURE

Sample preparation

Appropriate amounts of Bi_2O_3, TiO_2 and Na_2CO_3 powders were taken to produce bismuth titanate with a target stoichiometry of $Na_xBi_{5-x}Ti_4O_{15}$, where x = 0.25, 0.50 and 0.75 (corresponding to nominally 1, 2 and 3 at% Na). The powders were mixed in a ball mill for 4 hours with alumina grinding media and methanol as the dispersant. After ball milling the powders were dried at approximately 135°C for 4 hours. The powders were then screened to remove any large agglomerates and calcined in air at approximately 1100°C for 30 minutes in a covered alumina crucible heated in a batch-type furnace.

Polyvinyl alcohol (PVA) binder was then added and the powders pressed to produce compacts with a green density of approximately 4 g/cm^3. The pressed compacts were then dried at 75°C for about 24 hours. For bisque firing, the compacts were heated at a rate of 60°C/hour to 400°C and held at that temperature for 3-4 hours. The temperature was then increased at the rate of 150°C/hour to the final sintering temperature of 1100°C. Sintering was done in an oxygen rich atmosphere for 30 minutes in a covered alumina crucible. Thin discs (approximate thickness 0.5 mm) were then cut from the sintered compacts.

Heat treatment

To study the effect of heat treatment on the structural transformation, the powders were heated to a range of temperatures and held at the set temperatures for different lengths of time. The samples were furnace cooled at the end of the heat treatments**.

X-ray Diffraction of the samples was done using Cu Kα radiation and preliminary phase analysis done using the MAUD Rietveld refinement program [10] assuming isotropic strain. Based on these results, a high resolution X-ray scan of some of the samples was performed at the beamline X18A of the National Synchrotron Light Source, Brookhaven National Laboratory, NY, using 10 keV (wavelength 1.24Å) X-ray.

Transmission Electron Microscopy (TEM)

The samples were thinned to approximately 200 μm and discs of 3 mm diameter were cut from them. These were thinned further using a tripod polisher and finally placed on a copper grid and ion milled to make it electron transparent. High resolution TEM was done at Applied Materials' research lab, Santa Clara, CA using 200 kV accelerating voltage. To reduce the noise, some of the images were filtered by taking the Fourier transform, masking the spots and taking the inverse Fourier transform.

RESULTS

Structural transformations

X-ray diffraction of the powders revealed that the major phase in the calcined powders was $Bi_4Ti_3O_{12}$. X-ray diffraction of the sintered compacts using the Bruker AXS area detector diffractometer (Figure 2) show rings corresponding to the highest intensity peak – {109} plane of $Na_{0.5}Bi_{4.5}Ti_4O_{15}$ and spots corresponding to the {117} plane of the remaining $Bi_4Ti_3O_{12}$ phase indicating a transformation from $Bi_4Ti_3O_{12}$ to the $Na_{0.5}Bi_{4.5}Ti_4O_{15}$ phase. Heating the samples for 12 hours at 1050°C in air resulted in transformation to the cubic $Bi_2Ti_2O_7$ phase. This transformation from the $Bi_4Ti_3O_{12}$ structure was not observed in undoped bismuth titanate obtained from Cerac Corp., Milwaukee, WI.

Heat treatment

Synchrotron X-ray diffraction of the sodium-doped powders, which were heated to a range of temperatures from 700°C to 1100°C in intervals of 100°C, clearly indicated a transformation from the $Bi_4Ti_3O_{12}$ to the $Na_{0.5}Bi_{4.5}Ti_4O_{15}$ structure. A significant transformation was observed only in powders which were heated above 900°C (Figure 3). Rietveld refinement of the X-ray data indicates that the amount of $Bi_4Ti_3O_{12}$ structure decreases while $Na_{0.5}Bi_{4.5}Ti_4O_{15}$ structure increases.

High resolution TEM

TEM was done on the sample which showed the largest amount of $Na_{0.5}Bi_{8.5}Ti_7O_{27}$ phase. Figure 4 shows the image obtained using 200kV accelerating voltage. It is evident that the structures of $Bi_4Ti_3O_{12}$, $Na_{0.5}Bi_{4.5}Ti_4O_{15}$ and $Na_{0.5}Bi_{8.5}Ti_7O_{27}$ are closely related and a change in the stacking sequence of the layers can produce regions of different structures within each grain [11,12,13].

** Samples which were just ramped up to the set point (ramp rate 15 °C/min) and furnace cooled immediately are denoted as having a soak time of 0 minutes.

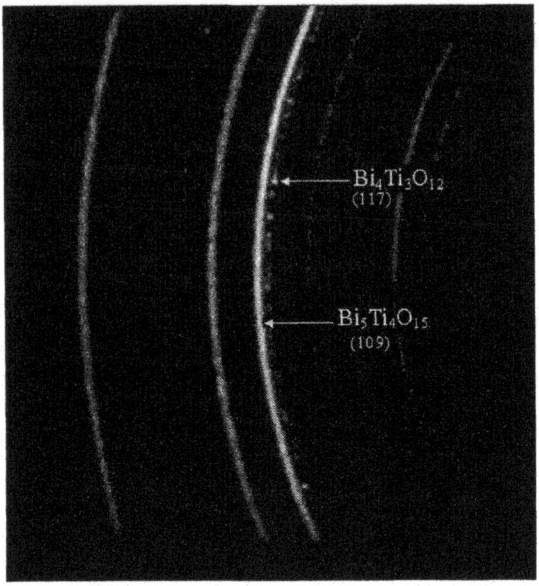

Figure 2: X-ray diffraction on an area detector diffractometer showing the major peak of the nominal $Bi_5Ti_4O_{15}$ phase and spots from the remaining $Bi_4Ti_3O_{12}$ phase. Note that the 2θ increases from right to left.

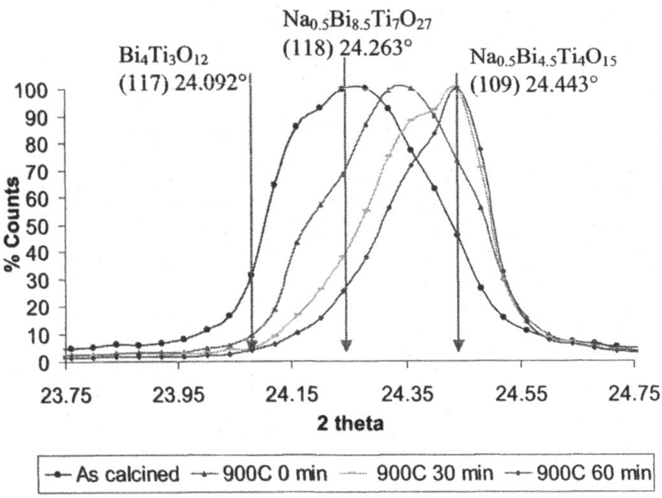

Figure 3: Shift of the highest intensity peak of bismuth titanate with soak time at 900°C.

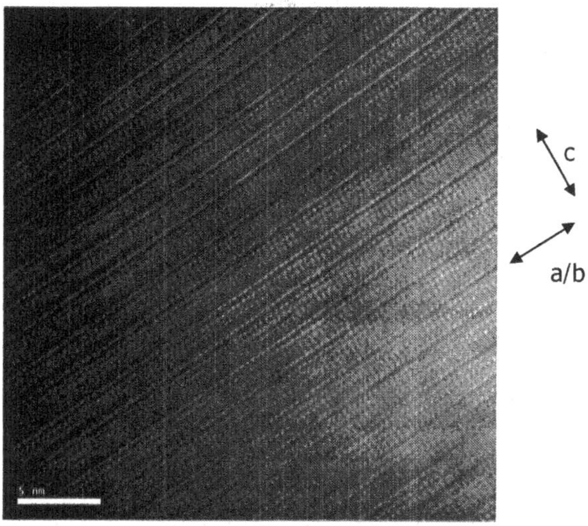

Figure 4: HRTEM showing the layer structure of bismuth titanate. Notice that the number of perovskite-like blocks between the $Bi_2O_2^{2+}$ layers varies across the grain

DISCUSSION

Formation of mixed-layer structures

The mixed-layer structure differs from the parent structures just in the number of perovskite-like blocks between the $Bi_2O_2^{2+}$ layers. Horiuchi *et al.* have studied the mixed-layer structure $Bi_7Ti_4NbO_{21}$ formed from the parent structures Bi_3TiNbO_9 and $Bi_4Ti_3O_{12}$. They have observed lattice defects in $Bi_7Ti_4NbO_{21}$ where the $Bi_2O_2^{2+}$ layer is displaced by one unit in the c-direction [12].

Since $Bi_7Ti_4NbO_{21}$ is itself a mixed-layer structure, the formation of these defects does not change the phase composition of the sample. But it can be reasoned that if similar defects form in the structurally uniform $Bi_4Ti_3O_{12}$ or Bi_3TiNbO_9 it can lead to the formation of a mixed layer structure. But in such a mixed-layer structure the number of perovskite blocks between the $Bi_2O_2^{2+}$ layers in the two parts of the mixed-layer structure differ by two (Figure 5). This type of defect in the stacking sequence cannot explain a mixed-layer structure in which the number of perovskite blocks in the two parts of the structure differs by one – as in the case of $Na_{0.5}Bi_{8.5}Ti_7O_{27}$.

Another possible mechanism by which the mixed-layer structure can form is by insertion of extra $Bi_2O_2^{2+}$ layers between slabs of perovskite-like blocks. This results in formation of dislocation-type defects [8, 11, 14]. The stress due to these defects can be accommodated either by slight adjustment in the width of the perovskite blocks close to the defect core or by side stepping of the $Bi_2O_2^{2+}$ layers [11]. Hutchison *et al.* have reported direct TEM evidence of the existence of both stress relaxation mechanisms [8, 11, 14]. Stress relaxation by side stepping of the $Bi_2O_2^{2+}$ layer by a single unit results in the perovskite slab on either side of the $Bi_2O_2^{2+}$ layer differing by a single layer of perovskite-like blocks (Figure 6). This is equivalent to producing the mixed-layer structure, such as $Na_{0.5}Bi_{8.5}Ti_7O_{27}$, where the number of perovskite blocks on either side of the central $Bi_2O_2^{2+}$ layer differs by one. The mobility of these defects increases on heating, resulting in a structural transformation as they move.

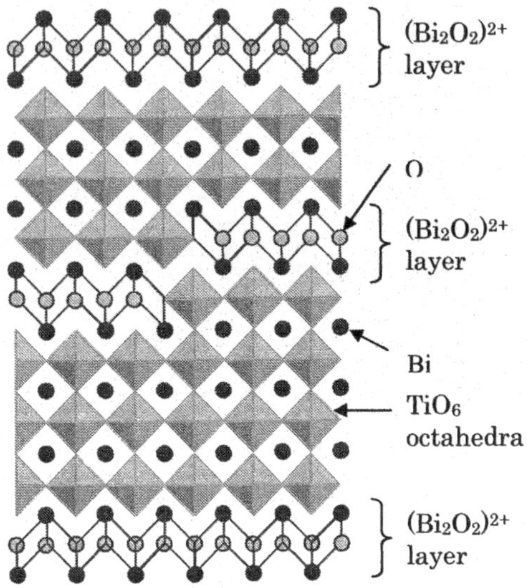

Figure 5: $Bi_2O_2^{2+}$ layer shifting in non-mixed structure giving rise to mixed-layer structure

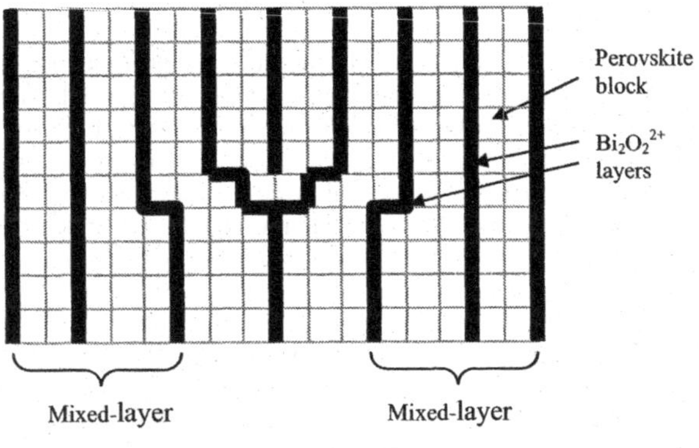

Figure 6:Schematic of the stress relaxation mechanism by side stepping of $Bi_2O_2^{2+}$ layers

CONCLUSIONS

A structural transformation on heating has been observed in sodium doped bismuth titanate. The transformation from the $Bi_4Ti_3O_{12}$ to the $Na_{0.5}Bi_{4.5}Ti_4O_{15}$ structure proceeds through the intermediate $Na_{0.5}Bi_{8.5}Ti_7O_{27}$ structure which has a mixed-layer type structure. HRTEM results indicate that the stacking sequence of layers in the mixed-layer type structure have a long range order. The transformation from the $Bi_4Ti_3O_{12}$ to the $Na_{0.5}Bi_{4.5}Ti_4O_{15}$ structure is facilitated by the movement of dislocation-type defects and stacking faults in the $Bi_2O_2^{2+}$ layers.

ACKNOWLEDGEMENTS

The authors wish to acknowledge the Keramos division, Piezo Technologies, Indianapolis, IN, for their assistance in sample preparation. We are also grateful to Hong Zhang of Applied Materials Inc., Santa Clara, CA., for assistance with TEM specimen preparation and high resolution imaging.

Financial support for this project was provided by the Indiana 21st Century fund. Research was carried out in part at the National Synchrotron Light Source, Brookhaven National Laboratory, which is supported by the U.S. Department of Energy, Division of Materials Sciences and Division of Chemical Sciences, under Contract No. DE-AC02-98CH10886.

REFERENCES

[1] T. Kikuchi, A. Watanabe, K. Uchida, "A Family of Mixed-Layer Type Bismuth Compounds," *Materials Research Bulletin*, **12** 299 (1977).

[2] E. C. Subbarao, "Crystal Chemistry of Mixed Bismuth Oxides With Layer-Type Structure," *Journal Of The American Ceramic Society*, **45** [4] 166 (1962).

[3] T. Kikuchi, "Synthesis of a New, Mix-Layered Bismuth Titanate $SrBi_8Ti_7O_{27}$," *Journal of Less-Common Metals*, **52** [1] 163 (1977).

[4] K. Uchida, T. Kikuchi, "Subsolidus Phase Equilibria in the System Na_2O-Bi_2O_3-TiO_2 at 1000°C," *Journal of the American Ceramic Society*, **61** [1-2] 5 (1978).

[5] J. L. Hutchison, F. J. Lincoln, J. S. Anderson, "Multiple Phases in The System Mgf_2-Nb_2O_5 An Electron Microscope Study Of Intergrowths, Defects, and Disordered Crystals," *Journal of Solid State Chemistry*, **10** 312 (1974).

[6] F. J. Lincoln, J. L. Hutchison, J. S. Anderson, "Electron Microscopy of the Niobium Oxides. Part II. Multiple Phases in the System $Nb_2O_5 + MgF_2$," *Journal of the Chemical Society (Dalton Transactions)*, 115 (1974)

[7] R. Gruehn, "Some Aspects of the Investigation of Intergrowth Phases in Nb_2O_5-Rich Systems," *N.B.S. Special Publication 364, Solid State Chemistry*, 63 (1972).

[8] J. L. Hutchison, J. S. Anderson, C. N. R. Rao, "Electron Microscopy of Ferroelectric Bismuth Oxides Containing Perovskite Layers," *Proceedings of the Royal Society of London. A.*, **355** 301 (1977).

[9] P.Boullay, G. Trolliard, D. Mercurio, J. M. Perez-Mato, L. Elcoro, "Towards a Unified Approach to the Crystal Chemistry of Aurivillius-Type Compounds - I. The Structure Model," *Journal of Solid State Chemistry*, **164** 252 (2002).

[10] L. Lutterotti, S. Matthies, H. R. Wenk, "MAUD (Material Analysis Using Diffraction): A User Friendly {Java} Program for{Rietveld} Texture Analysis and More," *Proceeding of the Twelfth International Conference on Textures of Materials (ICOTOM-12)*, **1** 1599 (1999).

[11] J. Hutchison, D. J. Smith, "High-Resolution Imaging of the Ferroelectric Perovskite $Ba_2Bi_4Ti_5O_{18}$," *Acta Crystallographica A*, **37** 119 (1981).

[12] S. Horiuchi, T. Kikuchi, M. Goto, "Structure Determination of a Mixed-Layer Bismuth Titanate, $Bi_7Ti_4NbO_{21}$, by Super-High-Resolution Electron Microscopy," *Acta Crystallographica A*, **33** 701 (1977).

[13] Y. Yoneda, J.Mizuki, R. Katayama, K. Yagi, H. Terauchi, S. Hamazaki, M. Takashige, "Stacking-Fault-Induced Intermediate Structure in Bismuth Titanate," *Applied Physics Letters*, **83** [2] 275 (2003).

[14] J. L. Hutchison, J. S. Anderson, C. N. R. Rao, "Lattice Images of Dislocations in the Ferroelectric Material $Ba_2Bi_4Ti_5O_{18}$," *Nature*, **255** 541 (1975).

DIELECTRIC PROPERTY OF OXYNITRIDE PEROVSKITES CONTAINING Ta^{5+}

Young-Il Kim and Patrick M. Woodward
Department of Chemistry, The Ohio State University
100 W 18th Ave., Columbus, Ohio 43210

ABSTRACT

Dielectric properties of perovskite oxynitrides, $A\mathrm{TaO_2N}$ (A = Ba, Sr, Ca), have been investigated by ac impedance spectroscopy using cold-isostatic-pressed polycrystalline powder specimens. Complex impedance spectroscopy is used to study dielectric properties of samples that cannot undergo an ordinary sintering process. Equivalent circuit analysis of impedance spectra showed that $BaTaO_2N$ and $SrTaO_2N$ are semiconducting ($\sigma_{el} \sim 10^{-5}\ \Omega^{-1}\mathrm{cm}^{-1}$) and have unexpectedly high bulk dielectric constants, $\kappa \sim 4900$ and 2900 respectively at room temperature. The dielectric constants of both compounds are frequency dependent and show a relatively weak, linear dependence upon temperature with no sign of a phase transition over the temperature range 300–180 K. However, $CaTaO_2N$ displayed distinctly different electrical behavior from the above two compounds: low dielectric constant ($\kappa < 20$) and much lower electrical conductivity ($\sigma_{el} \sim 10^{-8}\ \Omega^{-1}\mathrm{cm}^{-1}$). The tantalum oxynitride perovskites have optical band gaps corresponding to visible range in contrast to classical white dielectrics.

INTRODUCTION

From the 1930s, dielectric ceramics have been a topic of continual research interest owing to their use as essential components in various electronic devices.[1,2] As a representative example, recent technological developments in wireless telecommunication are based in part on the fabrication of superior dielectrics. Depending on the specific function of the product, different figures of merit are emphasized in evaluating dielectrics. These figures of merit are inevitably derived from fundamental dielectric properties: the dielectric constant (κ, relative permittivity), the dielectric loss ($\tan \delta$, dissipation), and the temperature dependence of these properties. In the case of high-density capacitor components, higher κ and lower $\tan \delta$ are desired for designing miniaturized and elegant devices, however, it is no less important that the manufactured components exhibit constant fidelity over a wide range of operational conditions. It is very difficult to simultaneously achieve high κ (>1000) and low temperature coefficient of dielectric constant (<10 ppm/K) due to the fact that the high relative permittivity of $\kappa > 1000$ is generally associated with a ferroelectric-paraelectric transition, and κ tends to be highly peaked at the transition temperature. This property of normal ferroelectrics seriously limits the useful temperature range of these materials in devices and has prompted the development of relaxor ferroelectrics.[3]

Recently we have investigated the dielectric properties of oxynitride perovskites.[4] Oxynitrides can be described as derivatives of oxides, formed by concurrent substitutions of cation and anion components, e. g. $SrTiO_3 \rightarrow SrTaO_2N$ ($Ti^{4+} + O^{-2} \rightarrow Ta^{5+} + N^{-3}$). This type of aliovalent substitution provides a mechanism for enhancing the dielectric polarizability through the substitution of more polarizable ions into the lattice. Partial replacement of O^{-2} with more negative and larger N^{-3}, needs to be compensated by introducing a cation substituent with higher valence and usually with smaller size. Clearly such evolutions in each component will increase the overall polarizability of the oxynitride crystal and eventually its κ.[5] Furthermore, the reduced electronegativity of the nitride ion, with respect to the oxide ion, will tend to increase the covalency of the cation-anion bonds, and in turn, increase the likelihood of cation displacements through a second-order Jahn-Teller distortion of the d^0 cation, thereby giving rise to a large static dipole. The mixed occupancy of the anion site in oxynitrides, $AM(O_{1-x}N_x)_3$, provides a similar condition to that found in relaxors, since the octahedral cations will experience random chemical environments in the absence of complete O/N ordering. It is therefore, an interesting matter to examine if the oxynitride perovskites possess intrinsically high κ and/or relaxor-like properties. Previously, Marchand *et al.* proposed that utilization of oxynitride dielectrics would have merit because the oxynitrides, stable in reducing atmosphere, could be fabricated with low cost non-jewelry metal electrodes.[6] However, relatively little work has resulted from this proposal. Herein we report the dielectric properties and electrical conductivities of four oxynitride perovskites, $BaTaO_2N$, $SrTaO_2N$, and $CaTaO_2N$ examined by impedance spectroscopy.

EXPERIMENTAL

Powder samples of perovskite oxynitrides, $ATaO_2N$ (A = Ba, Sr, Ca), were prepared by ammonolysis reaction. High-purity reagents of Ta_2O_5 and ACO_3 were mixed in an appropriate ratio using agate mortar and pestle, and heated in an alumina boat under flowing NH_3 gas (99.99 %) at 950–1000 °C for 80–100 h. For the $CaTaO_2N$, a mixture of KCl and NaCl (w/w 50:50) was used as a flux (mineralizer salt).

For the crystal structure determinations of final pure phases, X-ray powder diffraction data were collected using a Bruker D8 diffractometer (Cu Kα, λ = 1.5406 Å) and refined by the Rietveld method as incorporated in the GSAS software suite.[7] Dielectric properties were studied by the non-linear-least-squares fitting of impedance spectra measured using a commercial impedance analyzer (Solartron SI1260) between 10–10 MHz and over the temperature range 180–300 K. For the dielectric property measurement, pellet samples with ~45 % porosity were prepared by cold-isostatic-pressing at 60 kpsi, and the electrode contact was formed on both sides of pellet in the configuration of Ag/In-Ga/sample/In-Ga/Ag. For the equivalent circuit analysis of impedance spectra, a circuit model was constructed with two serial components each of which was taken as a resistor and a constant phase element in parallel.

RESULTS

All the three $ATaO_2N$ (A = Ba, Sr, Ca) structures could be prepared at temperatures ≤ 1000 °C, which is lower than that required for similar d^0 oxide perovskites, $AM^{+4}O_3$. In the ammonolysis reaction employed here, flowing NH_3 gas serves as nitriding agent and at the same time helps to lower the reaction temperature by providing a reducing atmosphere. Increasing the heating temperature did not accelerate the desired perovskite formation, since it simultaneously hastens the decomposition of NH_3 into N_2, which has negligible reactivity towards nitridation, and H_2. Heating of the ACO_3 + ½Ta_2O_5 mixtures in a flowing N_2 gas (99.99 %) resulted in corresponding muti-phase ternary oxides, indicating the overwhelming reactivity of oxygen even when present at trace levels.

Crystal structures of $ATaO_2N$ (A = Ba, Sr, Ca) were refined from X-ray powder diffraction patterns using the Rietveld method. The refinement results are given in Figure 1 along with a summary in Table I. It is found that the size of dodecahedral cation dictates the distortion and symmetry of the studied $ATaO_2N$ perovskites. In accordance to the calculated perovskite

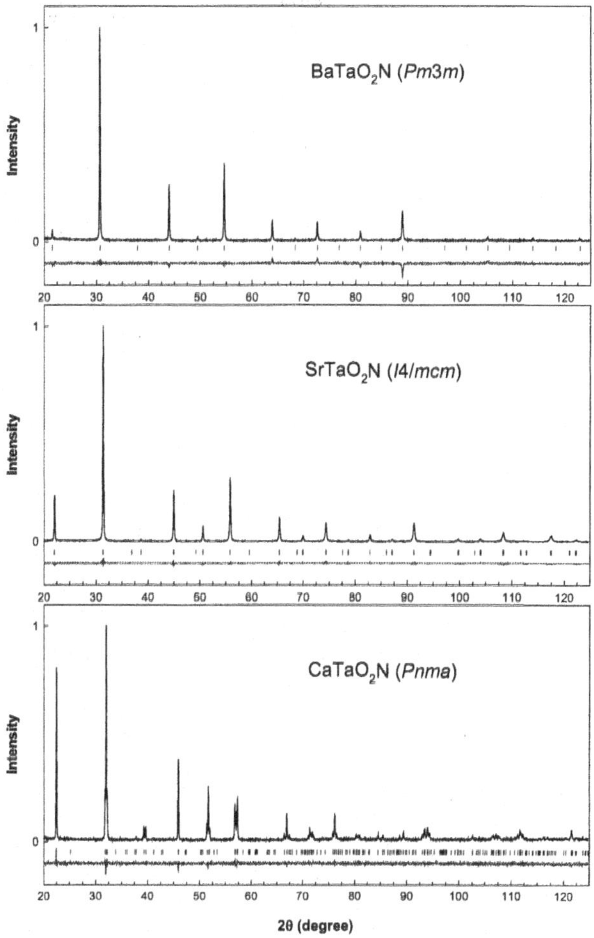

Figure 1. Rietveld refinement profiles for $ATaO_2N$ (A = Ba, Sr, Ca). Experimental data are presented together with Bragg reflection position markers and difference patterns.

tolerance factors (τ), cubic, tetragonal, and orthorhombic structures are observed for A = Ba (τ = 1.04), Sr (0.98), and Ca (0.95). The O/N ordering in these oxynitrides is an interesting problem relevant to the crystal and electronic structures and also to the dielectric behaviors, but X-ray powder diffraction is practically unable to distinguish oxide and nitride ions. Further examinations on the delicate structural features of $BaTaO_2N$ were conducted using high-resolution electron diffraction and neutron diffraction methods, but neither method revealed any indication that the crystallographic symmetry was different than the *Pm3m* symmetry of the ideal cubic perovskite structure. Since (Ta^{5+}–O^{-2}) and (Ta^{5+}–N^{-3}) would have distinct bond lengths it is reasonable to expect each single $Ta(O,N)_6$ octahedron is not of O_h symmetry, but if such distortions do take place cooperatively they must occur within sub-nanometer domains.

Table I. Summary of crystal structure refinements using X-ray powder diffraction patterns for $ATaO_2N$ (A = Ba, Sr, Ca) compared with that of $KTaO_3$ [Ref. 8].

	$BaTaO_2N$	$SrTaO_2N$	$CaTaO_2N$	$KTaO_3$
Space group	*Pm3m*	*I4/mcm*	*Pnma*	*Pm3m*
a (Å)	4.11250(5)	5.70251(6)	5.61895(4)	3.9883(2)
b (Å)		8.05420(16)	7.89309(5)	
c (Å)			5.54878(3)	
Volume (Å^3)	69.553(2)	261.912(3)	246.094(3)	63.44(1)
Z	1	4	4	1
$d_{Ta-(O,N)}$ (Å)	(6×) 2.05625(3)	(2×) 2.01355(4) (4×) 2.019(5)	(2×) 2.020(8) (2×) 2.026(2) (2×) 2.033(8)	(6×) 1.9942(1)
$\angle_{Ta-(O,N)-Ta}$ (°)	(3×) 180	(1×) 180 (2×) 173.8(5)	(1×) 153.9(4) (2×) 153.9(3)	(3×) 180

As-measured dielectric constants of $ATaO_2N$ (A = Ba, Sr, Ca) ceramic pellets are plotted in Figure 2 as functions of frequency (10^2–10^6 Hz) and temperature (180–300 K). It is noted that $BaTaO_2N$ and $SrTaO_2N$ possess high κ (> 500) which are frequency dependent and relatively invariant with the temperature, while $CaTaO_2N$ showed much lower κ (< 20) than the above two compounds over the entire frequency range. Examination of impedance spectra from each sample provides additional insight into the dielectric properties of the present ceramic samples. In contrast to the impedance spectra of typical dielectrics which are characterized as nearly-vertical lines converging to the origin (at the highest frequency), those of $BaTaO_2N$ and $SrTaO_2N$ are composed of two semi-circles as presented in Figure 3. The impedance spectra for $CaTaO_2N$ are rather featureless and not adequate for the detailed analysis.

A semi-circle on the impedance plane corresponds to an electrical component possessing finite resistance and capacitance. The simplified circuit representations of $BaTaO_2N$ and $SrTaO_2N$ should include two physical elements which are most likely to be bulk and grain boundary. Therefore the dielectric constants given in Figure 2 do not represent the intrinsic dielectric properties of $BaTaO_2N$ or $SrTaO_2N$ but the convolution of the bulk and grain boundary contributions in the ceramic pellets. It is not surprising to observe such significant grain boundary effects if we recall the high porosity (~45 %) of the pellet samples used. In order to separate the bulk and grain boundary properties in $BaTaO_2N$ or $SrTaO_2N$, their complex impedance spectra were analyzed using an equivalent circuit model shown in Figure 4, where the constant phase element (CPE) replaced the capacitor to account for slight depressions of semi-circles in impedance plots. As displayed also in Figure 4, reliable non-linear-least-squares fittings were obtained for all the data sets, where the dielectric constants for bulk parts could be extracted. The dielectric constants of $BaTaO_2N$ and $SrTaO_2N$, as obtained from equivalent circuit analysis, are shown in Figure 5. Both compounds are found to possess very high κ values

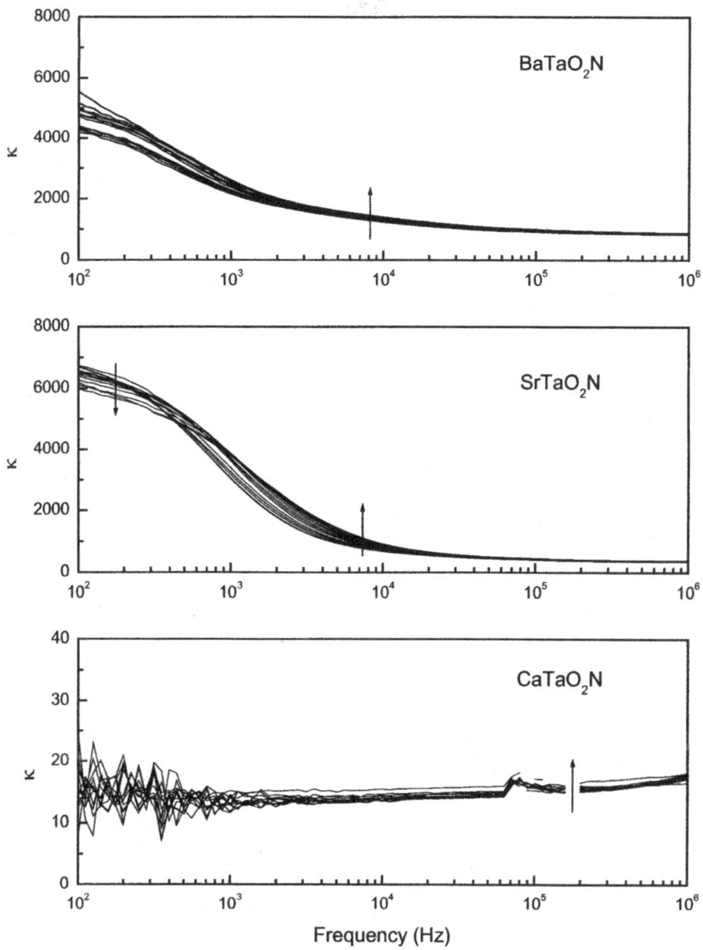

Figure 2. Frequency-dependent dielectric constant of ATaO$_2$N (A = Ba, Sr, Ca) measured at 180–300 K. Arrows indicate the increasing direction of temperature. The data represent the combined bulk and grain boundary contributions within each ceramic sample.

exceeding 2500 at the temperature range 180–300 K. It is noteworthy that such high κ's of present compounds show only moderate temperature dependences with temperature coefficients of +1140 ppm/K and −730 ppm/K for BaTaO$_2$N and SrTaO$_2$N, respectively. Thermally activated electrical conductivities of ~10^{-5} Ω^{-1}cm^{-1} (180–300 K) were found from bulk BaTaO$_2$N and

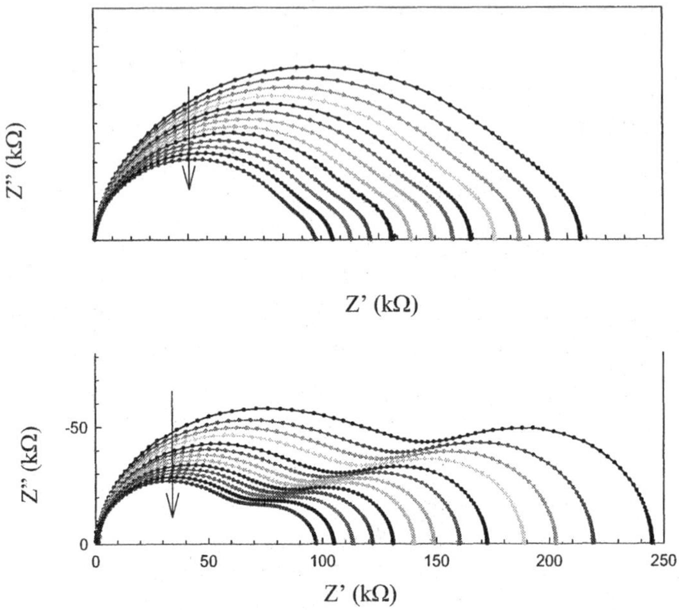

Figure 3. Temperature-dependent impedance spectra for (top) $BaTaO_2N$ and (bottom) $SrTaO_2N$ measured at 1–10^7 Hz and at 180–300 K with a 10 K step. Arrows indicate the increasing direction of temperature.

$SrTaO_2N$, also by the equivalent circuit analysis. The electrical conductivities of present oxynitrides are several orders higher than those of typical oxides containing Ta^{5+}, which can be associated with the optical band gap modulations. Compared with the related perovskite oxides, the oxynitrides in this study have smaller band gaps ($KTaO_3$, 3.5 eV, $BaTiO_3$, 3.4 eV, $BaTaO_2N$, 1.8 eV, $SrTaO_2N$, 2.1 eV) owing to the N 2*p* contribution to the top of valence band. As deduced from the electrical conduction behavior, dielectric losses of $BaTaO_2N$ and $SrTaO_2N$ were measured to be quite high, ~20 % at 1 MHz. The defect concentration from the imperfect sintering should be in part responsible for the high level of dielectric loss.

Dielectric properties of crystalline solids can often be understood through accurate crystal structure information from features such as static ion displacements and order-disorder behavior. The *Pm*3*m* symmetry of $BaTaO_2N$, determined by neutron[9], electron[4], and X-ray diffraction studies is seemingly inconsistent with the observed high dielectric constants. We believe that the apparent contradiction can be traced to the fact that the above methods give information on the average structure and fail to take into account local structural distortions. In fact a preliminary extended X-ray absorption fine structure (EXAFS) spectroscopy at the Ta L_3-edge revealed a significant deformation of first coordination shell around Ta, suggestive of the polar displacements within individual $Ta(O,N)_6$ octahedron. Nonetheless, the absence of superstructure

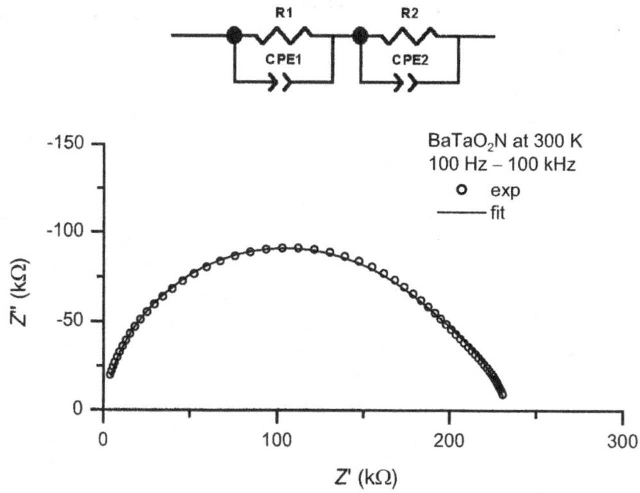

Figure 4. (Top) Model used for the equivalent circuit analysis for $BaTaO_2N$ and $SrTaO_2N$ and ceramic pellets and (bottom) the impedance fit for $BaTaO_2N$ at 300 K.

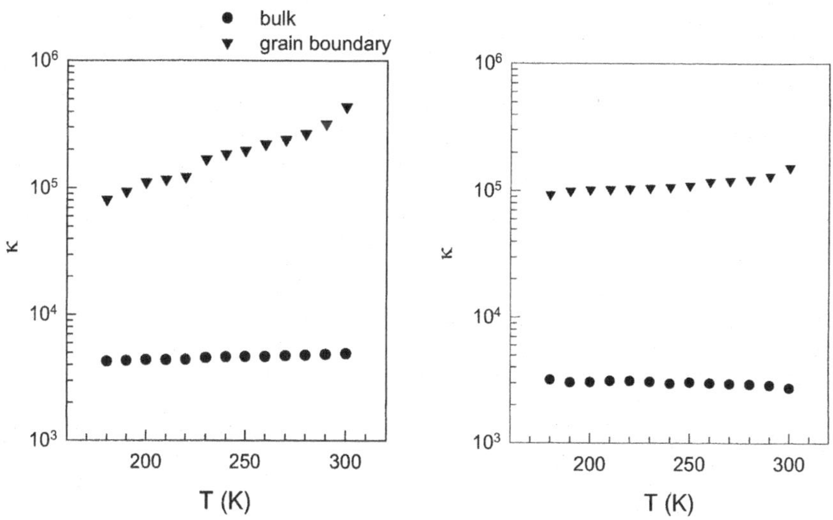

Figure 5. Dielectric constants of (left) $BaTaO_2N$ and (right) $SrTaO_2N$. Bulk and grain boundary components are separately evaluated.

reflections in diffraction experiments can only be rationalized if the coherence length of these displacements is extremely small. Such a picture is consistent with a disordered anion arrangement. This disorder will inhibit phase transitions and explain the fairly small temperature dependence of κ.

In the $SrTaO_2N$, a subtle tetragonal distortion exists resulting in *I4/mcm* symmetry, but a comparable dielectric behavior is observed to that in cubic $BaTaO_2N$. However the $CaTaO_2N$, with a more pronounced distortion to orthorhombic *Pnma* symmetry and larger octahedral tilting, exhibits a markedly different electrical property from those of Ba- and Sr- analogues. It can be conjectured that the drastic distinction of $CaTaO_2N$ arises from the severe structural distortion of perovskite lattice which suppresses out-of-center displacement of Ta and any cooperative polarization phenomena between adjacent octahedral units. Nevertheless, at this time we cannot rule out the possibility that the dramatically different dielectric behaviors of $BaTaO_2N$ and $CaTaO_2N$ are due in part to differences in the degree of sintering at grain boundaries. Studies of thin film samples are currently in progress to gain a better understanding of the intrinsic properties of these materials. Finally, we note that the initial characterization suggests that $SrTaO_2N$ and $BaTaO_2N$ have dielectric constants with very similar magnitudes but with opposite signs of temperature coefficients, leaving a possibility to realize a near-zero temperature coefficient from $Ba_{1-x}Sr_xTaO_2N$ solid solution.

CONCLUSION

The oxynitride perovskites, $BaTaO_2N$ and $SrTaO_2N$, are shown to be interesting dielectric materials possessing high relative permittivity with weak and linear temperature dependences. Normal ferroelectric behavior is not observed due to disorder in the anion distribution, which prevents long-range cooperative ion displacements. Diffraction analyses using neutron, electron, and X-ray could not reveal any sign of superstructure or symmetry lowering in $BaTaO_2N$. Instead it appears as though local geometric distortions occur with very small coherence lengths.

REFERENCES

[1]P.K. Davies, R.S. Roth, Chemistry of Electronic Ceramic Materials, National Institute of Standards and Technology Special Publication 804 NIST Gaithergburg, MD 1991.

[2]R.Roy, R. Guo, A.S. Bhalla, “Perovskite: lessons from its history and its crystal chemistry”, *Ceramic Transactions*, Vol. 104, Amer. Ceram. Soc. Westerville, OH (1999)

[3]L.E. Cross, “Relaxor ferroelectrics” *Ferroelectrics*, **76** 241 (1987).

[4]Y.-I. Kim, P.M. Woodward, K.Z. Baba-Kishi, and C.W. Tai “Characterization of the structural, optical, and dielectric properties of oxynitride perovskites AMO_2N (A = Ba, Sr, Ca; M = Ta, Nb)” *Chemistry of Materials*, **16** 1267 (2004).

[5]R.D. Shannon “Dielectric polarizabilities of ions in oxides and fluorides” *Journal of Applied Physics*, **73** 348 (1993).

[6]R. Marchand, F. Pors, Y. Laurent, O. Regreny, J. Lostec, J.M. Haussonne,“Oxynitride perovskites used as dielectric materials” *Journal de Physique*, **C1** 901 (1986).

[7]A.C. Larson, R.B. von Dreele General Structure Analysis System, Los Alamos National Laboratory Report LAUR 86-748 (1994).

[8]E.A. Zhurova, Y. Ivanov, V. Zavodnik, V. Tsirelson, “Electron density and atomic displacements in $KTaO_3$” *Acta Crystallographica*, **B56** 594 (2000).

[9]Y.-I. Kim, P.M. Woodward, T.Vogt, unpublished result.

HIGH ENERGY DENSITY FERROELECTRIC GLASS-CERAMICS

Ming-Jen Pan
Nova Research, Inc.
c/o Code 6350, Naval Research Laboratory
4555 Overlook Avenue SW
Washington, DC 20375

Michael Lanagan
Materials Research Institute
The Pennsylvania State University
University Park, PA 16802

Barry A. Bender
Code 6350, Naval Research Laboratory
4555 Overlook Avenue SW
Washington, DC 20375

Ching-Tai Cheng
Materials Research Laboratories
Industrial Technology Research Institute
Hsinchu, Taiwan 31040

ABSTRACT

In this study we examined the feasibility of using glass-ceramics with ferroelectric precipitates as high energy density dielectric materials. Specific goals were to demonstrate: 1) the formability of ferroelectric particles inside a glass matrix, 2) high dielectric constant and breakdown strength of glass-ceramics, and 3) improved energy density. Three glasses with network former (Si, Al, and B) content as low as 20 wt% was successfully formed using splash-quenching. After the crystallization temperature of each composition was determined by differential thermal analysis, a series of heat treatment were performed. X-ray diffraction results confirmed the formation of the appropriate ferroelectric phase (barium titanate for two compositions and barium strontium niobate for the third). The dielectric constant of the heat-treated samples can be up to 300 and the dielectric breakdown strength up to 800 kV/cm. Correspondingly, the projected energy density is 6-8 joules/c.c., an order of magnitude higher than commercial ceramic dielectrics.

INTRODUCTION

The discovery of barium titanate in the 1940's was a major event in the development of electronic ceramics. In the past few decades, however, the improvements in the properties of dielectric ceramics have been marginal. Most of the efforts were focused on the engineering of multilayer manufacturing and the implementation of barium titanate-based dielectrics for the consumer market. There were only occasional reports on the development of high energy density dielectrics such as the lead lanthanum zirconate titanate (PLZT) family of ceramics whose energy density can be as high as 12 joules/cc [1].

One should take note that it is unrealistic to expect the energy density of dielectric materials to reach hundreds of joules per cc. From the physics point of view, the energy density of capacitors will never reach that of electrochemical energy storage devices. Applying an electric field to a dielectric material causes polarization to develop by electrical displacement, which is analogous to merely stretching the molecular bonds of the dielectric. In contrast, the molecular bonds inside a battery are actually severed and then rearranged to achieve a lower energy state through chemical reaction. The trade-off is the high rate of energy release, or power density, in capacitors. A qualitative comparison of existing energy storage devices is shown in Figure 1 [2].

When we examine the energy density of various dielectrics, it was clear that the storage capacity of ceramic dielectrics is limited by the low dielectric breakdown strength, which largely is due to the presence of porosity that acts as the source of (electrical) stress concentration. In the past various techniques have been used to remove the porosity from a ceramic body during sintering, including vacuum sintering and hot isostatic pressing. However, none of these techniques could completely remove the existing pores due to thermodynamically unfavorable boundary conditions in the final stage of sintering. In contrast, in the glass-ceramic approach, we start with a porosity-free glass and then precipitate ferroelectric particles inside the glass matrix. Because the glass-ceramic after heat treatment is free of porosity, its breakdown strength may approach its intrinsic value of a few MV/cm, an order of magnitude higher than conventional ceramic dielectrics [3]. Certainly the presence of a low K glass phase imposes a tradeoff in the final energy density, but the effect can be minimized by employing a high volume fraction of high K ferroelectric particles. A schematic of the glass-ceramic processing is shown in Figure 2.

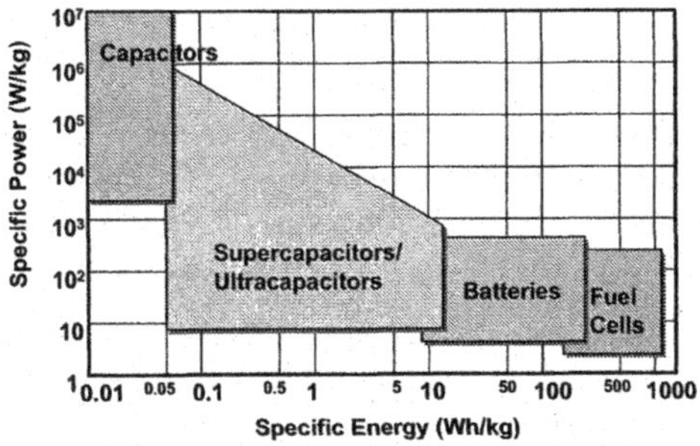

Figure 1 A comparison of the various energy storage technologies (adapted from [2])

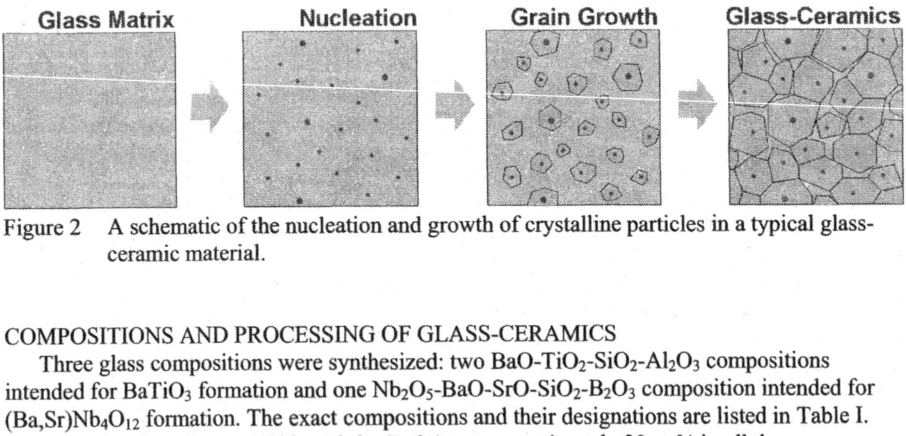

Figure 2 A schematic of the nucleation and growth of crystalline particles in a typical glass-ceramic material.

COMPOSITIONS AND PROCESSING OF GLASS-CERAMICS

Three glass compositions were synthesized: two BaO-TiO_2-SiO_2-Al_2O_3 compositions intended for $BaTiO_3$ formation and one Nb_2O_5-BaO-SrO-SiO_2-B_2O_3 composition intended for $(Ba,Sr)Nb_4O_{12}$ formation. The exact compositions and their designations are listed in Table I. Note that the glass formers (SiO_2, Al_2O_3, B_2O_3) are approximately 20 wt% in all three compositions in order to maximize the amount of ferroelectric constituents and hence to increase the dielectric constant of the crystallized glass-ceramics.

The oxide and carbonate starting materials were mixed and then melted at 1500°C for 2 hours. The glass melt was quenched on an aluminum slab to form a thin sheet of glass. The rapid cooling prevents spontaneous devitrification. The quenched glass was immediately placed in a 600°C furnace for annealing.

Table I. The composition of the three glass-ceramic candidates (in wt%)

Material	Weight %		
	BT1	BT2	BT3
BaO	36.7	36.7	19.4
TiO_2	32.4	32.4	--
SiO_2	14.2	14.2	17.1
Al_2O_3	7.7	7.7	1.15
F_2	0.6	--	--
MgO	1.3	1.3	--
Excess BaO	12.7	13.3	--
Nb_2O_5	--	--	19.55
PbO	--	--	10.2
B_2O_3	--	--	5.8
SrO	--	--	8.8

CRYSTALLIZATION BEHAVIOR

To determine the crystallization temperature, differential thermal analysis (DTA) was performed on crushed glass samples. All three compositions exhibited a strong exothermic peak and the corresponding crystallization temperatures are ~720°C for BT1 and BT2, and 755°C for NBO, as shown in Figure 3. The glass samples were heat-treated at temperatures equal to or higher than the crystallization temperature to explore the kinetics of crystallization. The heat treatment profiles are listed in Table II.

After heat treatment, the samples transformed from transparent to opaque, indicating that the crystallization process occurred. X-ray diffraction (XRD) patterns clearly showed the transformation from an amorphous state to a crystalline state (Figure 4). The presence of crystalline phases was detectable at temperatures as low as 800°C. The primary crystalline phase was identified as barium titanate in BT1 and BT2 samples, and $(Ba,Sr)Nb_4O_{12}$ in NBO samples.

Table II. The heat treatment schedule used in this study

	Nucleation Condition	Grain Growth Condition	
		Temp	Dwell
1	10°C/min ramp	800°C	2 hrs
2	10°C/min ramp	900°C	2 hrs
3	10°C/min ramp	1000°C	2 hrs
4	10°C/min ramp	1000°C	12 hrs
5	10°C/min ramp	1000°C	48 hrs
6	10°C/min ramp	1000°C	120 hrs
7	720°C/0.5 hrs*	1000°C	2 hrs
8	720°C/10 hrs*	1000°C	2 hrs

*720°C for BT1 and BT2; 720°C for NBO

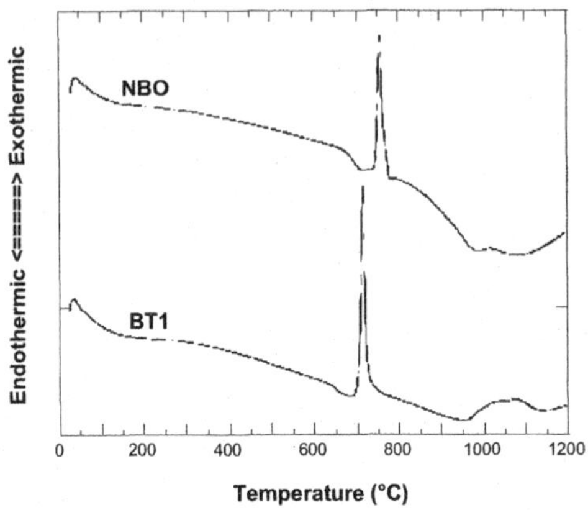

Figure 3 DTA curves were used to determine the crystallization temperature. Note that BT1 and BT2 curves are almost identical (only one is shown).

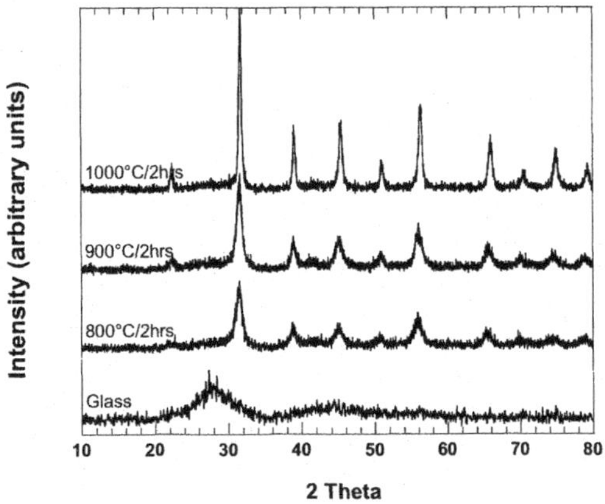

Figure 4 X-ray diffraction pattern of BT2 showing the development of the $BaTiO_3$ crystalline phase after heat treatment.

The density values of BT1, BT2, and NBO before and after heat treatment are 4.46/5.10, 4.43/5.10, and 5.07/5.20 g/cm3, respectively. This was expected as the ordered crystalline phase always has a higher density than its amorphous phase of the same composition. We also noticed that there was little variation in density among samples with different thermal histories, suggesting that the crystallization probably take place rapidly in a narrow temperature range.

The fracture surface of heat-treated samples was examined using a scanning electron microscope (SEM). In a number of samples, ferroelectric particles were too small (<100 nm) to be observed on SEM, although their presence was confirmed by XRD results. In general, samples treated at 1000°C for more than 10 hours have larger ceramic grains and (>100 nm) exhibited distinctive features similar to those of typical ceramic samples (Figure 5). No apparent porosity was observed.

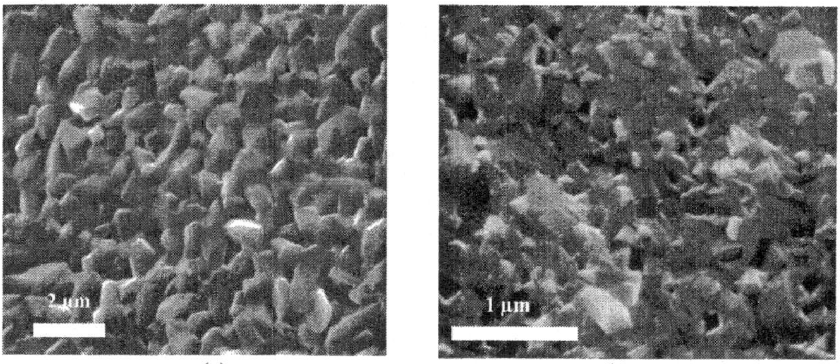

(a) (b)

Figure 5 Fracture surface of BT2 (a) and NBO (b) heat-treated at 1000°C for 48 hours.

DIELECTRIC CONSTANT AND BREAKDOWN STRENGTH

The dielectric constant of heat-treated samples is shown in Figure 6 as a function of temperature and thermal history. Note that BT1 and BT2 showed almost identical behavior so only BT1 were plotted. In general, the dielectric constant increased with prolonged heat treatment as the crystallinity of the precipitated particles became more developed. The maximum dielectric constant of all three compositions is in the range of 250 to 300, while the temperature dependence is within the X7R range. We also examined the effect of dwell time at the crystallization temperature on the final dielectric properties. Figure 7 shows that there was little variation in dielectric constant among samples with different dwell times, suggesting that the crystallization occurred rapidly such that the subsequent dwell became irrelevant.

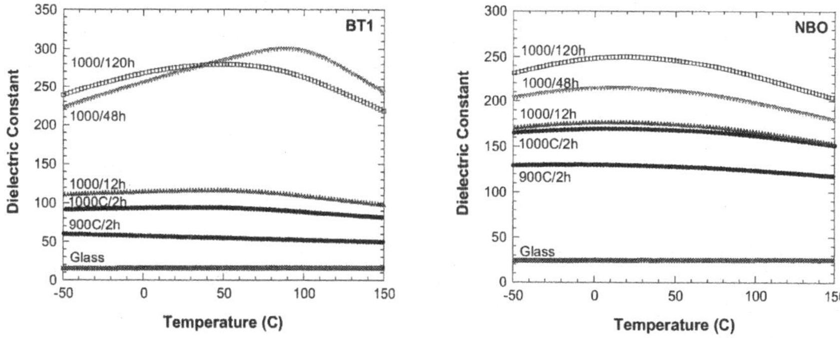

Figure 6 The dielectric constant of glass-ceramics as a function of thermal history. Note that BT1 and BT2 showed almost identical behavior.

Several NBO samples were prepared for the breakdown strength measurement. The test was performed using a Hi-Pot tester (QuadTech Guardian 12KVDC) with a ramp rate of 200 V/sec until breakdown occurred. The results are tabulated in Table III. The breakdown strength is several times higher than conventional ceramic dielectrics (typically <100 kV/cm). We also tested the breakdown strength of several commercially available glass-ceramics, all of which are low-K ceramic substrate materials. We discovered that the average breakdown strength of Dupont 951 material is 1900 kV/cm and that of Heraeus CT-2000 is 1400 kV/cm. These values approach those of polymer dielectrics.

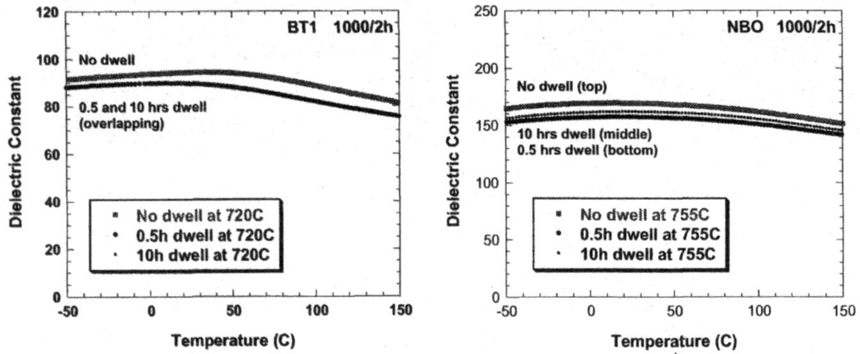

Figure 7 The dielectric constant of BT1 and NBO treated at 1000°C for 2 hours with different dwell time at the nucleation temperature.

Table III. Dielectric breakdown strengths of NBO samples

	Dielectric Thickness (μm)	Breakdown Voltage (kV)	Breakdown E-Field (kV/cm)
900°C/2hrs	~120	6.5 – 8.0	500 – 700
1000°C/2hrs	120 – 150	5 – 10	400 – 800

PROJECTED ENERGY DENSITY

Due to the lack of appropriate samples in the final stage of the feasibility study, a direct energy density measurement using a discharge circuit was not possible. Nevertheless, if the combination of dielectric constant of 300 and breakdown strength of 750 kV/cm is used to *project* the energy density, one can get approximately 6 to 8 joules/cc (Figure 8). As this is only an exploratory study and many improvements can be made in the future, we anticipate further improvement in the energy density,possibly reaching 20-30 joules/cc.

CONCLUSIONS

We demonstrated the feasibility of using ferroelectric glass-ceramic materials as high energy density dielectrics. In this study, the candidate compositions exhibited maximum dielectric constant of approximately 250-300 and the dielectric breakdown strength in the range of 400-800 kV/cm, which is 5 to 10 times higher than that of conventional ceramic dielectrics. The results projected a final energy density of 6 to 8 joules/cc. We believe that energy density higher than 10 J/cc is achievable (possibly reaching 20-30 J/cc) by refining the glass-ceramic compositions and processing parameters.

Despite the successful results, there are a number of issues to be addressed in the near future. First is the determination of energy density by using a discharge circuit. The concern is whether the dielectric constant of the glass-ceramics shows appreciable decrease under high electric field, which has been observed in barium titanate-based dielectrics [3]. The effects of residual stress will also be examined. As the crystalline phase has higher density than the glass phase, there may be considerable residual stress present after precipitation. It is not clear how the residual stress would affect the dielectric properties and breakdown strength. Finally, the kinetics of ceramic particle precipitation is being studied and the results will be published elsewhere in the near future.

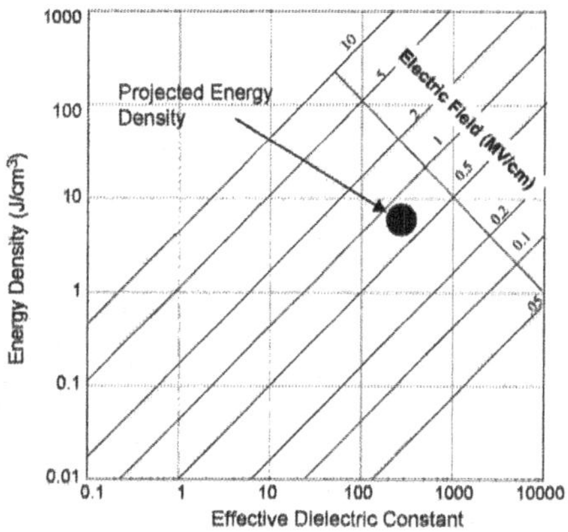

Figure 8 The projected energy density of ferroelectric glass-ceramics

REFERENCES

1. K. wa Gachigi, *Electrical Energy Storage in Antiferro-electric-Ferroelectric Phase Switching, Chemically Modified Lead Zirconate Titanate Ceramics*, Ph.D. Thesis, The Pennsylvania State University, 1997.
2. R. Kötz and M. Carlen, "Principles and Applications of Electrochemical Capacitors," *Electrochimica Acta* **45**, p2483-2498 (2000)
3. G.R. Love, "Energy Storage in Ceramic Dielectrics," *J. Am. Ceram. Soc.*, **73** [2] p323-28 (1990).

INCONGRUENT VAPORIZATION OF $AgNbO_3$

L. P. Cook, W. Wong-Ng, and Z. Yang
Ceramics Division
National Institute of Standards and Technology
Gaithersburg, MD 20899 USA

ABSTRACT

The high-temperature behavior of $AgNbO_3$ is an important aspect of its processing for applications as a dielectric material. We have studied the $AgNbO_3$ decomposition reaction by simultaneous differential thermal analysis/thermogravimetric analysis (DTA/TGA) at p_{total} = 0.1 MPa, and by vacuum thermogravimetry at p_{total} = 2.7 mPa. The DTA/TGA results indicate a decomposition temperature of ≈ 860 °C at p_{O2} = 0.1 kPa. From x-ray analysis of the products of vacuum thermogravimetry experiments over the range 650 °C to 800 °C, we conclude that the decomposition reaction is:

$$52\ AgNbO_3(s) \rightarrow 2\ Ag_8Nb_{26}O_{69}(s) + 36\ Ag(s) + 9\ O_2(g).$$

The incongruent vaporization of $AgNbO_3$ must be taken into account for applications in low-temperature co-fired ceramic (LTCC) materials, typically processed at 800 °C to 900 °C, if low oxygen pressures are employed during processing.

1. INTRODUCTION

$AgNbO_3$ is of current interest as the end member of a series of solid solutions with high dielectric permittivity and reduced temperature dependence of permittivity [1, 2]. Potential uses of doped $AgNbO_3$ include lead-free piezoelectric devices for medical and defense applications, and materials for advanced microwave applications [3, 4]. Many of these applications will involve high-temperature ceramic processing, or packaging via an LTCC route, in which thick film technology is used to laminate dielectric components and associated micro-circuitry into a compact, robust, relatively low-cost product. The temperature stability of the dielectric components is of obvious importance, as processing temperatures are in the 800 °C to 900 °C range and above. Furthermore, the effect of both internal and ambient p_{O2} during processing must be taken into account. The goal of our research on $AgNbO_3$ is to establish its thermal stability at various p_{O2}, and ultimately to determine its relationship to the thermochemistry of other phases in the Ag-Bi-Nb-O system, a system important as a model for interfacial interactions in LTCC and other multi-layered materials [5, 6].

2. CRYSTALLOGRAPHY OF $AgNbO_3$

Figures 1 and 2 show the structure of $AgNbO_3$ along the a- and c-axis respectively. At room temperature this phase is orthorhombic [7], with a space group *Pbcm*, a=5.5462(3) Å, b=5.6028(4) Å, and c=15.6365 (13) Å. The structure basically consists of a 3-dimensional network of corner-sharing distorted NbO_6 octahedra. These octahedra are tilted and rotated with respect to each other.

$AgNbO_3$ undergoes complex phase transformations as a function of temperature, with up to six transitions reported [8-12], although not all these transitions are well-characterized. According to Kania and Kwapuliniski [8], x-ray diffraction [13], electron diffraction [14], DTA, and dielectric domain structure studies [15, 16] have showed that in $AgNbO_3$ ceramics, the following sequence of phase transformations takes place as a function of temperature:

67 °C:	O (M1) (F) → O(M2) (A)
267 °C:	O (M2) (A) → O(M3) (A)
353 °C:	O (M3) (A) → O(O1) (P)
361 °C:	O(O1) (P) → O(O2) (P)
387 °C:	O(O2) (P) → T (P)
579 °C:	T (P) → C (P),

where

M1, M2, M3	= orthorhombic symmetry in rhombic orientation
O1, O2	= orthorhombic symmetry in parallel orientation
F	= ferroelectric
T	= tetragonal
C	= cubic
A	= antiferroelectric
P	= paraelectric.

While the high temperature phase transitions (M3→O1, O1→O2, O2→T and T→ C) are detectable by x-ray, DTA, dielectric and domain structure investigations, the low temperature phase transitions (M1→M2 and M2→M3) are observed only as diffuse maxima of the permittivity function and cannot be detected by x-ray methods [17]. Although these complex phase transitions may have some influence on the stability and properties of $AgNbO_3$ at lower temperatures, it is significant that above 579 °C, $AgNbO_3$ has the classic cubic perovskite structure. No additional solid state phase transitions have been reported above this temperature.

3. EXPERIMENTAL METHODS[1]

Stoichiometric $AgNbO_3$ was synthesized from reagent-grade Ag_2O and Nb_2O_5. The powders were weighed, homogenized in a mortar and pestle, pressed into a pellet, and fired at 750 °C in air for 7 h, re-homogenized and fired at 850 °C in air for 4 days. X-ray powder diffraction analysis (XRD) of the product showed it to be single-phase (Fig. 3).

DTA/TGA studies were completed on a Mettler TA1 thermoanalyzer equipped with Anatech digital control firmware and data collection software. The temperature ramp rate was 10 °C/min. The thermocouple was calibrated against the α/β quartz transition (571 °C), and the melting points of NaCl (801 °C) and Au (1064 °C). Reported temperatures have an uncertainty of ± 5 °C (standard error of estimate). Samples were contained in dense, slip-cast MgO crucibles. During the DTA/TGA experiments, gas of known oxygen content, supplied from analyzed Ar/O_2 mixtures, was flowed through the

[1] Certain commercial equipment, instruments, or materials are identified in this paper to foster understanding. Such identification does not imply recommendation or endorsement by the National Institute of Standards and Technology, nor does it imply that the materials or equipment identified are necessarily the best available for the purpose.

sample area. During progressive heating, as the sample became reduced, the evolved oxygen increased the local p_{O2} around the sample, thereby retarding further reduction, and leading to a broadening of the DTA signal. The TGA signal was less affected, and by using the 1st derivative of the TGA signal, a better indication of the onset of the decomposition reaction could be obtained. This approach was tested on reduction of CuO to Cu_2O at p_{O2} = 0.1 kPa. By noting the temperature at which the 1st derivative of the TGA signal made a measurable progression from the linear baseline, a CuO onset decomposition temperature of 800 °C was determined (Fig. 4), in excellent agreement with the literature value of 796 °C [18]. Consequently, this method was used to determine the solid state decomposition temperature of $AgNbO_3$.

Because the products produced by decomposition in the conventional DTA/TGA method were subject to back-reaction while the samples cooled down, a different approach was used to determine the solid state decomposition products of $AgNbO_3$. In this approach, $AgNbO_3$ was loaded into an MgO Knudsen cell with a 0.5 mm orifice [19], placed in the thermoanalyzer, evacuated, and slowly decomposed at 650 °C to 800 °C. The decomposition reaction was monitored thermogravimetrically. During these experiments, the background pressure was maintained at ≈ 2.7 mPa. When the experiments were terminated, the vacuum was maintained as the samples cooled, eliminating the possibility of re-oxidation.

Starting materials and products in this study were characterized by powder XRD using a Philips 2θ diffractometer with Cu K_α radiation and a graphite monochromator. Diffractometer control and data acquisition were achieved using the JADE[tm] software system.

4. RESULTS AND DISCUSSION

As preliminary calculations indicated that the vapor pressure of Ag was not a factor over the temperature range of interest [18], it was assumed that all mass loss observed thermogravimetrically (after taking into account initial outgassing of the sample and buoyancy effects) was associated with oxygen loss. Initial experiments at various p_{O2} indicated that $AgNbO_3$ may melt without undergoing solid state decomposition. For example, in pure oxygen, the melting point is 1133 °C (Fig. 5); there was no thermogravimetric indication of a major oxygen loss corresponding to a decomposition event below this temperature in 100 % O_2. Similarly, DTA/TGA events corresponding to solid state decomposition by oxygen loss were not observed at p_{O2} = 7.5 kPa. However, solid state decomposition was observed when the p_{O2} was considerably lower, at 0.1 kPa (Fig. 6). At p_{O2} = 0.1 kPa, analysis of the 1st derivative of the TGA data in Fig. 6 suggests an $AgNbO_3$ decomposition onset temperature of ≈ 860 °C. To assess the nature of the solid state decomposition reaction of $AgNbO_3$, products of the vacuum decomposition experiments on $AgNbO_3$, as described above, were analyzed by powder XRD (Fig. 7). Results showed metallic Ag, as expected, but the remainder of the peaks in Fig. 7 could not be explained by any combination of the reported (Ag, Nb)-oxide phases, including $Ag_2Nb_4O_{11}$ and $Ag_2Nb_8O_{21}$ [20]. However, it was observed that the remainder of the peaks in Fig. 7 are topologically identical to the pattern reported for $Ag_8(Nb_{0.5}Ta_{0.5})_{26}O_{69}$ [20]. After taking into account the shift in x-ray parameters expected to result from substitution of Nb for Ta, the residual peaks of Fig. 7 fit the expected pattern of a proposed $Ag_8Nb_{26}O_{69}$ analog very well. To test this possibility, a sample with the $Ag_8Nb_{26}O_{69}$ stoichiometry was prepared, and calcined at 800 °C in air for 5 days. The products gave a powder XRD pattern having only the peaks of the proposed analog (upper pattern in Fig. 7). We therefore conclude that an end-member phase, an analog of $Ag_8(Nb_{0.5}Ta_{0.5})_{26}O_{69}$, may exist, with the corresponding stoichiometry $Ag_8Nb_{26}O_{69}$. Further experiments on the structure and stability of the proposed phase are in progress. Most importantly, the XRD results suggest that the $AgNbO_3$ decomposition reaction can be written as:

$$52\ AgNbO_3(s) \rightarrow 2\ Ag_8Nb_{26}O_{69}(s) + 36\ Ag(s) + 9\ O_2(g) \qquad (1).$$

The dependence of reaction (1) on p_{O2} is undergoing further investigation by Knudsen effusion vacuum thermogravimetry.

5. SUMMARY AND CONCLUSIONS

We have shown that the perovskite phase, $AgNbO_3$, a potential dielectric material, undergoes vacuum decomposition by incongruent vaporization at 650 °C to 800 °C to produce metallic Ag and a phase postulated as $Ag_8Nb_{26}O_{69}$. At p_{O2} = 0.1 kPa, the $AgNbO_3$ decomposition temperature is ≈ 860 °C. At higher p_{O2}, $AgNbO_3$ appears to melt at a higher temperature without solid state decomposition. At lower p_{O2}, it is expected on a thermodynamic basis that the decomposition temperature will be lower. The typical LTCC processing window is 800 °C to 900 °C. Therefore, if $AgNbO_3$ dielectrics are included in an LTCC laminate, oxygen-dependent effects might occur during processing: a) if processing occurs at low p_{O2} (< 0.1 kPa) for very short times, the $AgNbO_3$ may locally control the internal p_{O2} of the laminate by partial decomposition according to reaction (1), thereby functioning as an internal oxygen buffer; b) if processing occurs at low p_{O2} for an extended time, the $AgNbO_3$ may completely decompose according to reaction (1), as the oxygen diffuses out of the ceramic or is consumed by neighboring phases; c) if processing occurs at high p_{O2} (> 7.5 kPa), our DTA/TGA data suggest that $AgNbO_3$ will be stable with regard to reductive decomposition up to its melting point. Processing of $AgNbO_3$ ceramics will benefit from a more complete knowledge of the dependence of decomposition temperature on p_{O2}, and of the kinetics of the decomposition reaction.

ACKNOWLEDGEMENTS

We gratefully acknowledge partial financial support for this work from the NIST-MSEL Director's Reserve Fund.

REFERENCES

[1] M. Valant and D. Suvarov, "New High-Permittivity $AgNb_{1-x}Ta_xO_3$ Microwave Ceramics: Part II, Dielectric Characteristics," *Journal of the American Ceramic Society*, **82** [1] 88-93 (1999) Valant and Suvarov (1999).

[2] M. Valant, D. Suvorov, C. Hoffmann, and H. Sommariva, "Ag(Nb,Ta)O_3-Based Ceramics with Suppressed Temperature Dependence of Permittivity", *Journal of the European Ceramic Society*, **21** [15] 2647-2651 (2001).

[3] I. Grinberg and A. M. Rappe, "Silver Solid Solution Piezoelectrics," *Applied Physics Letters*, **85** [10] 1760-1762 (2004).

[4] H. T. Kim, T. Shrout, C. Randall and M. Lanagan, "Low-Temperature Sintering and Dielectric Properties of $Ag(Nb,Ta)O_3$ Composite Ceramics," *Journal of the American Ceramic Society* **85** [11] 2738-2744 (2002).

[5] L.P. Cook, W. Wong-Ng, P. Schenck, M. Vaudin and J. Suh, "A Model System for Interfacial Reactions in LTCC Materials," pp. 177-182 in "Proceedings of IMAPS International Microelectronics and Packaging Society (IMAPS) Conference on Ceramic Interconnect Technology, April 7-9, 2003, Denver, CO," (2003).

[6] L.P. Cook, W. Wong-Ng, and J. Suh, "Interfacial Reactions in LTCC Materials," in C.L. Claeys, W. Wong-Ng, and K.M. Nair, editors, "Low-Temperature Electronics and Low-temperature Cofired Ceramics-Based Electronic Devices", Electrochemical Society/American Ceramic Society (in press).

[7] C. N. W. Darlington, "An X-ray Diffraction Study of $AgNbO_3$ and Comparison with $NaNbO_3$," *Powder Diffraction* **14** 251-257 (1999).

[8] A. Kania and J. Kwapulinski, "$Ag_{1-x}Na_xNbO_3$ (ANN) Solid Solutions: from Disordered Antiferroelectric $AgNbO_3$ to Normal Antiferroelectric $NaNbO_3$," *Journal of Physics – Condensed Matter*, **11** 8933-8946 (1999).

[9] S. P. Solov'ev, Yu. N. Venevtsev, and G. S. Zhdanov, "An X-ray Study of Phase Transitions in $NaNbO_3$," *Soviet Physics – Crystallography*, **6** [2] 171-175 (1961).

[10] C. N. W. Darlington and H. D. Megaw, "The Low-Temperature Phase Transition of Sodium Niobate and the Structure of the Low-Temperature Phase, *N*," *Acta Crystallographica*, **B29** 2171 (1973).

[11] C. N. W. Darlington and K. S. Knight, "High-Temperature Phases of $NaNbO_3$ and $NaTaO_3$," *Acta Crystallographica*, **B55** 24-30 (1999).

[12] I. Lefkowitz, K. Lukaszewica and H. D. Megaw, "The High-Temperature Phases of Sodium Niobate and the Nature of Transitions in Pseudosymmetric Structures," *Acta Crystallographica*, **20** 670-683 (1966).

[13] M. Pawelczyk, "Phase-Transitions in $AgTa_xNb_{1-x}O_3$ Solid-Solutions," *Phase Transitions*, **8** [4] 273-292 (1987).

[14] M. Werwerft, D. Van Dyck, V. A. M. Brabers, J. Van Landayt, and S. Amelinckx, "Electron-Microscopic Study of the Phase-Transformations in $AgNbO_3$," *Physica Status Solidi A – Applied Research*, **112** [2] 451-466 (1989).

[15] A. Kania, "$AgNb_{1-x}Ta_xO_3$ Solid Solutions – Dielectric Properties and Phase Transitions," *Phase Transitions* **3** 131-139 (1983).

[16] A. Kania, "An Additional Phase Transition in Silver Niobate $AgNbO_3$," Ferroelectrics, **205** 19-28 (1998).

[17] S. Miga and J. Dec, "Reorientation of the W' Domain Walls in Ferroelastic Silver Niobate Crystal," *Journal of Applied Physics*, **85** [3] 1756-17459 (1999).

[18] C.W. Bale, P. Chartrand, S.A. Degterov, K. Hack, R. Ben Mahfoud, J. Melancon, A. D. Pelton, and S. Petersen, "FactSage Thermochemical Software and Databases," *CalPhad*, **26** 189-228 (2002).

[19] L. P. Cook, R. Klein, and W. Wong-Ng, "Thermodynamics of MgB_2 – by Calorimetry and Knudsen Thermogravimetry," *IEEE Transactions - Applied Superconductivity*, accepted for publication (2005).

[20] Powder Diffraction File (PDF), produced by ICDD-International Centre for Diffraction Data, 12 Campus Blvd., Newtown Square, PA 19073.

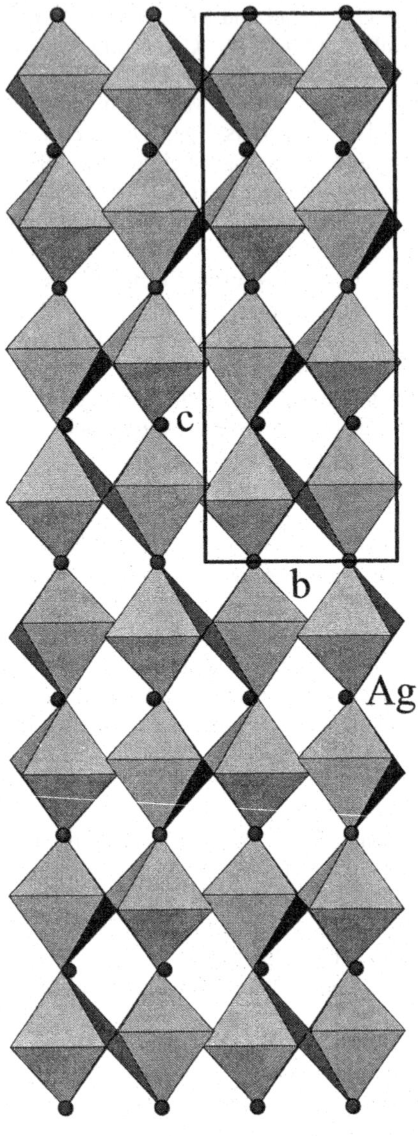

Fig. 1. Structure of $AgNbO_3$ viewing along the *a*-axis. The structure consists of distorted NbO_6 octahedra. Ag atoms are represented as spheres.

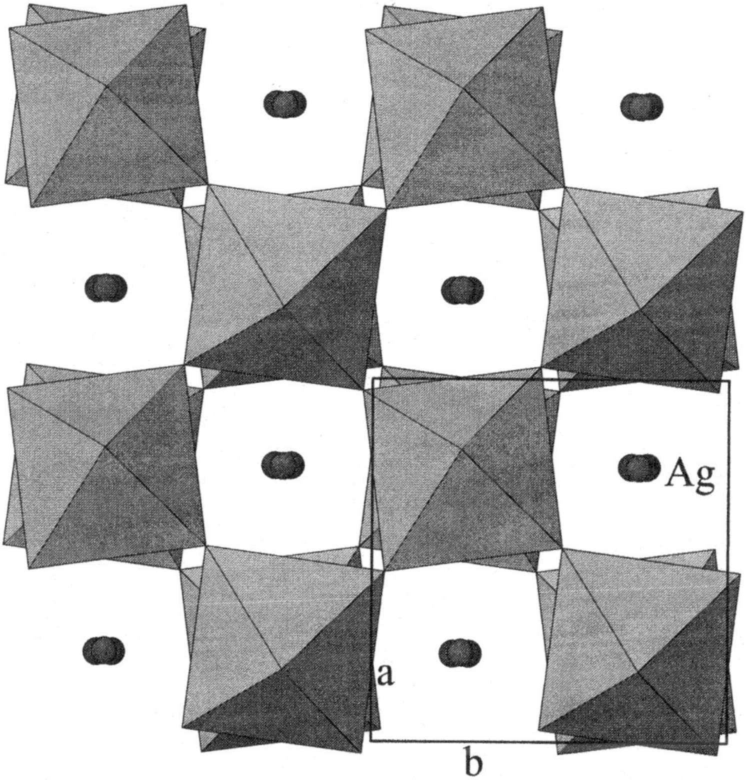

Fig. 2. Structure of $AgNbO_3$ viewing along the *c*-axis. The structure consists of distorted NbO_6 octahedra. Ag atoms are represented as spheres.

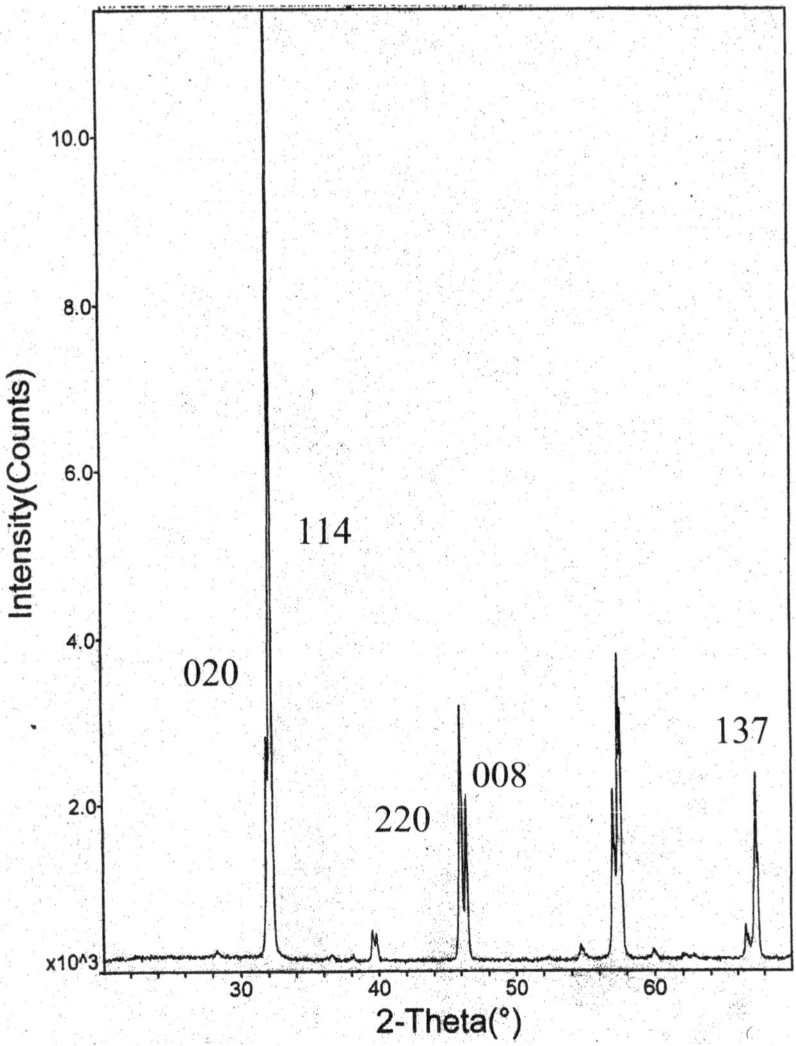

Fig. 3. Powder x-ray diffraction pattern of $AgNbO_3$ synthesized for this study

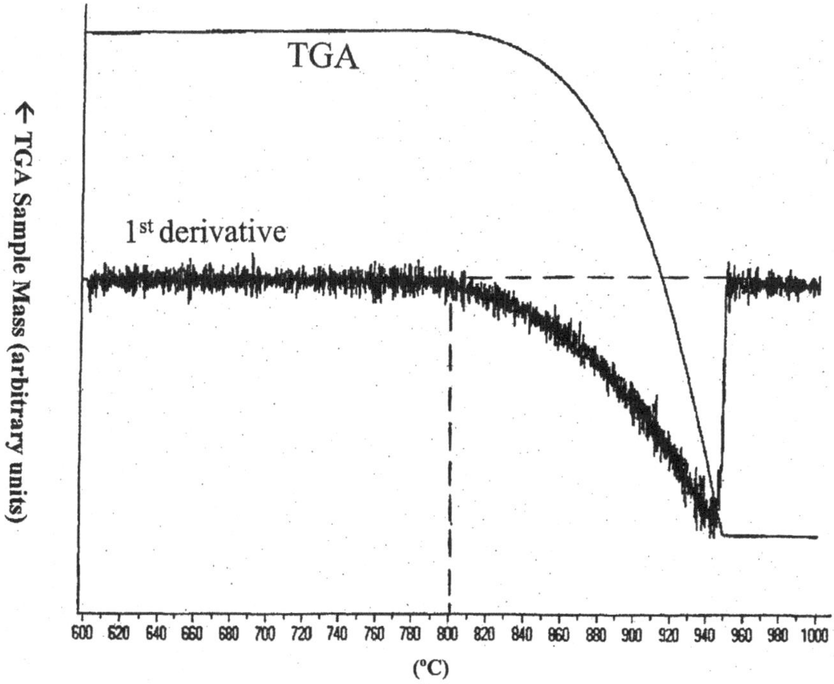

Fig. 4. TGA data for CuO/Cu_2O reaction

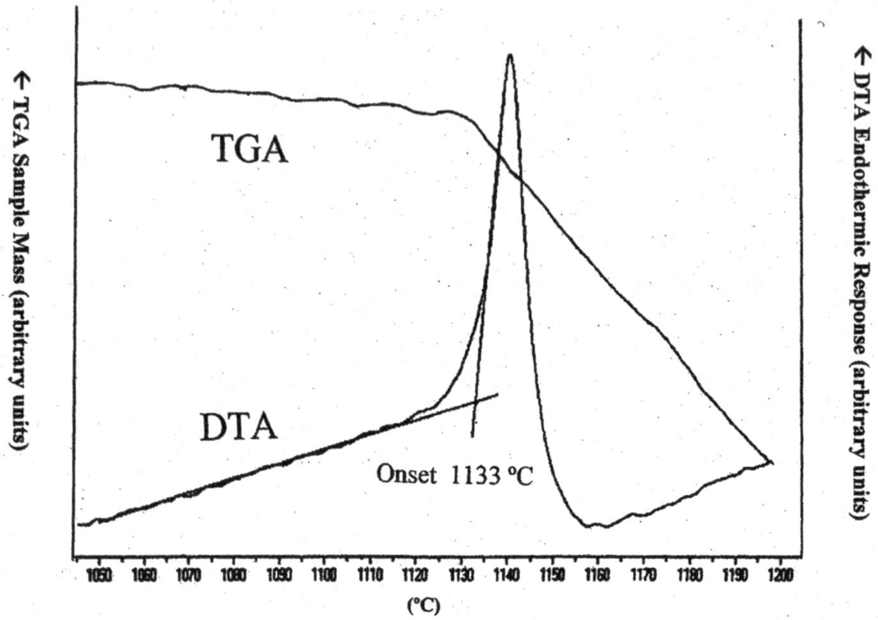

Fig. 5. DTA/TGA data for melting of $AgNbO_3$

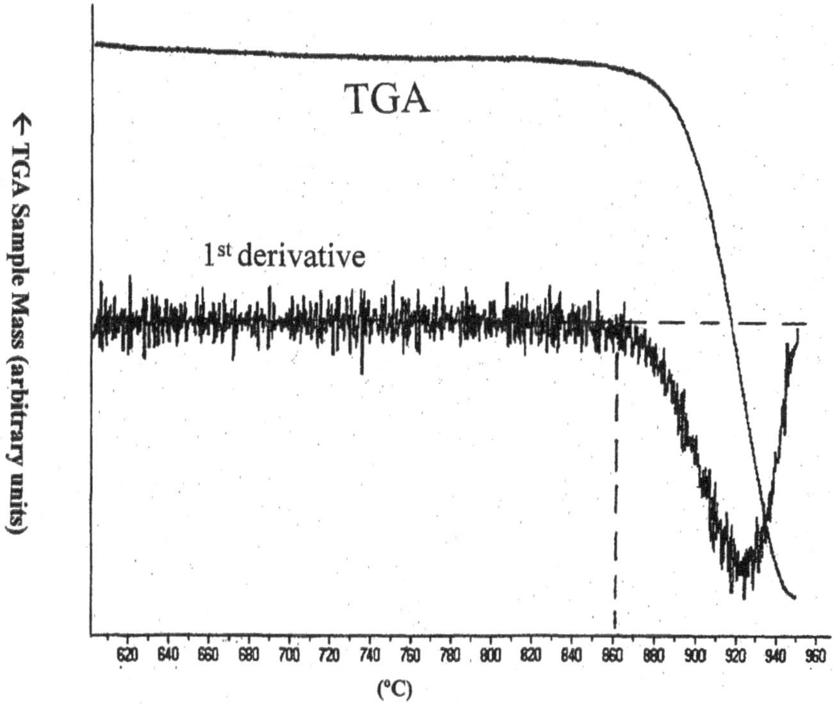

Fig. 6. TGA data for decomposition of $AgNbO_3$ at $pO_2 = 0.1$ kPa

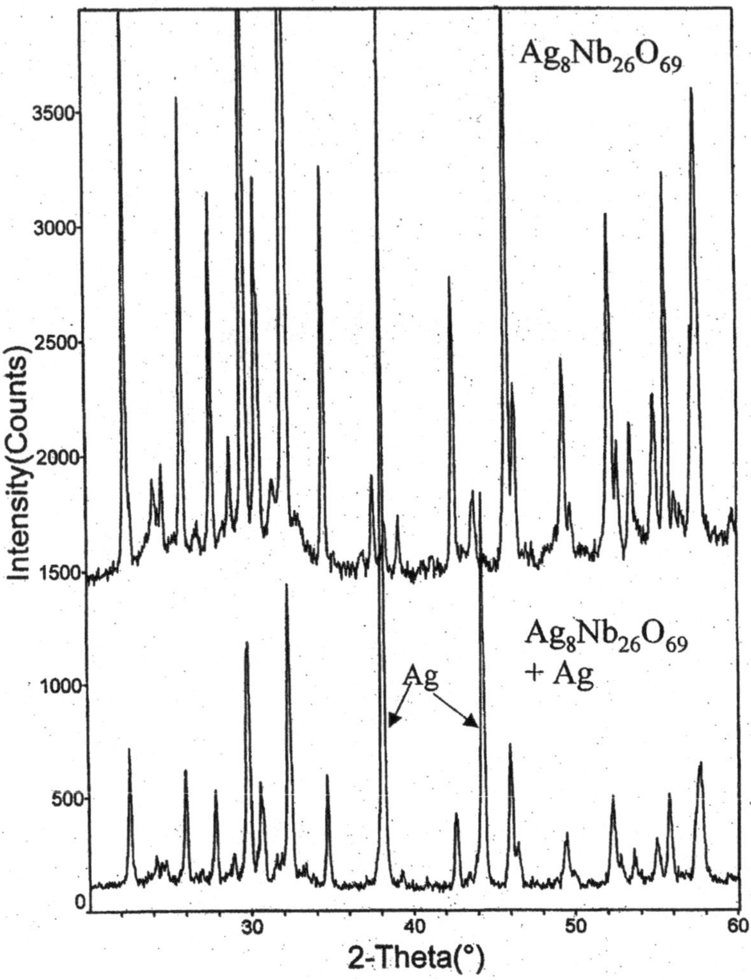

Fig. 7. Powder x-ray diffraction pattern of $AgNbO_3$ vacuum decomposition product (lower pattern) compared with pattern of annealed $Ag_8Nb_{26}O_{69}$ composition (upper pattern)

Author Index

Keyword Index

Printed and bound by CPI Group (UK) Ltd, Croydon, CR0 4YY

15/09/2026

14962410-0001